DATE DUE

Return of
the Crazy Bird

Return of the Crazy Bird

The Sad, Strange Tale of the Dodo

Clara Pinto-Correia

COPERNICUS BOOKS

An Imprint of Springer-Verlag

Published in the United States by Copernicus Books,
an imprint of Springer-Verlag New York, Inc.
A member of BertelsmannSpringer Science+Business Media GmbH

Copernicus Books
37 East 7th Street
New York, NY 10003
www.copernicusbooks.com

Library of Congress Cataloging-in-Publication Data
Correia, Clara Pinto.
   Return of the crazy bird: the sad, strange tale of the dodo / Clara Pinto-Correia.
      p. cm.
   Includes bibliographical references (p.) and index.
   ISBN 0-387-98876-9 (alk. paper)
      1. Dodo.   I. Title.
   QL696.C67   C67   2002
   598.6'5—dc21   2002070737

Manufactured in the United States of America.
Printed on acid-free paper.

9 8 7 6 5 4 3 2 1

ISBN 0-387-98876-9   SPIN 10730283

In the fondest memory of Steve Gould—
may his wisdom keep guiding me.

Contents

Preface

I caught my first glimpse of the legendary dodo when, as a child, I heard an amazing tale about this incredible bird.

I can't remember how old I was. Still, I'm sure this revelation occurred during one of those long conversations between my parents and their friends as they sat around the living room talking one evening. I loved these gatherings and would listen with rapt attention to their cascades of words and sentences that always seemed so loaded with erudition and wit. As I listened, I would hold my breath and pray to Jesus to make me able to shine the way they did when I was older. It was a feat that seemed absolutely out of reach, and yet it was probably the thing I desired most as a child. In order to learn their language, I would listen very closely and try to memorize expressions, bits of trivia, big words full of syllables, awesome stories.

So, one day as I was listening, someone said that the infamously stupid and now extinct dodo bird owed its name to my native Portuguese, who had been the first Europeans to find the bird's island home. Portuguese sailors had promptly called the beast a *doudo*, the ancient version of our modern *doido*, an idiot, a fool, or any creature out of its mind. Then, with time and different waves of colonization, *doudo* had ended up as *dodo*. As a child, I was even more of a rabid, shameless patriot than I am now, so my lungs took a big breath of nationalistic pride, and I never forgot this precious piece of information.

Around the time I turned ten, one of my many cousins, the studious one who wanted to become a teacher, offered me *Alice's Adventures in Wonderland*. *Alice*, it turned out, proved to be much harder reading than the other books I read at the time. It was different, and I couldn't really figure it out. It seemed funny to me, for instance, that a young girl falling into a bottomless well would take the time to wonder how much latitude and longitude she could have covered. I guessed right from the start that something much grander, something much more wild and complex was going on beneath the surface—only I couldn't grasp it and ended up feeling quite frustrated. One of my parents' friends, a math professor, told me that Lewis Carroll was the pen name of one Charles Dodgson, a tutor of logic and mathematics at Christ Church College in Oxford, and then proceeded to explain some of the mathematical curiosities in the book.

Although I wasn't all that interested in the story behind the story at the time, I now understand that the problem with *Alice* and its sequel *Through the Looking Glass* is that they are books for adults, and then some. Only adults with a sound scientific background can appreciate all of the scientific riddles that Carroll wove into the story. Alice's changes of scale—her growth or her shrinking depending on what she happened to have just drunk or just eaten—are anticipations of Lorentz's transformations, which describe time's expansions or space's contractions as derived from relativity. Everything in Carroll's books is relative, as masterfully exemplified by the conversation between the four participants in the mad tea scene: "Do I see what I eat, or do I eat what I see?" "Do I breathe as I sleep, or do I sleep as I breathe?" Even at the beginning, as Alice is falling slowly into the rabbit's hole, the narrator observes that "Either the well was very deep or she fell very slowly." Later, when the young girl finds a seemingly useless key in a hallway lined with locked doors, the dilemma is the same: "Either the locks were too large or the key was too small."[1]

And then there are all the puns, all those plays with words and their meanings, as in the mouse's *tale* becoming the mouse's *tail*—certainly very long, as Alice promptly notices. Ludwig Wittgenstein, who was a philosopher as well as an architect, had some affinities with Carroll: they both cared for the relationship between what a thing *is* and what a thing *is called*. *Mesa* doesn't obviously have the same meaning as *table*. Or does it? Many of the puzzles and scientific quirks in Carroll's book address this question.

For a child oblivious to the riddles in the book, *Alice* is full of weird creatures. Some come from decks of cards, such as the Queen of Hearts, who wants everybody beheaded. The White Rabbit spends his time entering and exiting the scene in great agitation. And then there is the Mad Hatter, who is the result of a chemical accident. I learned much later that the Hatter's madness must have been the consequence of poisoning with mercury, since mercury's salts were then routinely used to create the shiny felt of hats. And then there were all these strange animals never seen in our backyards, such as the griffin and the walrus. And, of course, the dodo.

At that time, I thought that the dodo had been invented by the Portuguese navigators and was, just like the griffin, a mythological creature. There were no dodos in my grandmother's cookbooks or in the illustrated zoology magazines for children I got from school libraries or grabbed with delight in dentists' waiting rooms.

Having spent the formative years of my childhood in Angola (then a Portuguese colony), I had decided early on that I wanted to become a park ranger. With time, as the dream kept becoming more polished, I was going to be the next Jane Goodall. I was still dreaming of a future along these lines when I started studying biology at 18. While none of these dreams came true in the end, biological studies allowed me to discover the real story of the once-real dodo by the end of the 1970s. The dodo became a great metaphor that I used to explain the deeper, scientific meaning of ecology to the philistine majority at a time when the word "ecosystem" was generally equated with naked pot-smoking hippies with flowers in their hair. The story was also straightforward and simple. And it was a historical fact—not a tale by the Grimm brothers after they had a chat with Aesop.

As I could now explain, thanks to my fledgling biological knowledge, there once had been a huge, heavy, flightless bird in Mauritius. It was perfectly adapted to the island's peaceful life. It lived on nuts cracked open with its powerful beak, the only other tool needed for its survival besides a pair of robust legs. Because Mauritius did not have any predators, the dodo knew nothing about fear and self-defense. It was, as experts would put it today, "ecologically naïve." Then, the dodo's peaceful universe was shattered by consecutive waves of Europeans who began to arrive on the shores of Mauritius in the early sixteenth century. The explorers killed the poor bird by the hundreds with only sticks, stones, or even their bare hands. At the same time, they also released all kinds of smart, hungry, omnivorous pests and pets, such as rats, pigs, dogs, cats, and even monkeys. These creatures competed with the dodo for natural resources, stole dodo eggs from their undefended nests, attacked the adults, and destroyed the bird's natural habitat. Consequently, the dodo was "discovered" and destroyed in less than a century. Its swift disappearance marked the first time in history where the intervention of man clearly triggered an animal species' extinction. It was an impressive story, and I told people this tale as often as I could. For some 20 years now, I've been meaning to write a book to teach readers everything about the dodo. Little did I know how much I was going to have to teach myself in the process.

I would like you to take this book as a book of many things at once. Like the word, *dodology*, it is about everything having to do with the dodo. I like to believe that my text also works as a detective story—a hint here, a piece of evidence there, a broken thread recovered later, a net closing in on the reader as the mystery draws to a close. I also intended it to be a rosary of exemplary stories overlapping one another in the fashion of *Arabian Nights*—"it happened, o fortunate reader," and here we go in yet another direction.

Clara Pinto-Correia
Lisbon, Portugal, August 2002

———————————

1 Lewis Carroll and John Tenniel (illus.), *Alice's Adventures in Wonderland; and, Through the Looking-Glass and What Alice Found There*, New York: Hurst, 1903.

Acknowledgments

To the team of the John Carter Brown Library at Brown University for their hospitality and the marvelous arsenal of voyage books they introduced me to.

To the team at the Fine Arts Library of Harvard University, for their stupendous collection and friendly guidance to the strange world of Rudolf II and his painters.

To Christoph Luethy, for his precious help, insights, translation jobs and detective work as he helped me discover the face and story of the most elusive Roelandt Savery.

To Bob Richards and Paula Findlen, for their most helpful suggestions.

To Fernando Mascarenhas, descendant of Pedro Mascarenhas, for his personal search in his private library to help me discover the man who gave the Mascarenes their name.

To my fellow Portuguese novelist Luisa Costa Gomes, for her great and merciless editing skills and unconditional sense of friendship.

To Tim Yohn, the editor who got this story off the ground, for great moments and sheer intellectual fun.

To Anna Painter, for her saintly patience.

To Stephen Jay Gould, as always.

To Dick, Joseph, Mike, and Ricky, for having so graciously put up with me and my laptop during all those long months when it often seemed that the dodo was more important to me than my own family.

Chapter 1

The Weirdest Creatures

WITHIN THE WESTERN TRADITION, maps of the world have often been populated with strange creatures. In the first century A.D., Pliny the Elder wrote *Natural History*, filling the lands conquered by the Roman Empire with animals both real and mythological. There were the amphisbaenia, snakes with a head at each end that were associated with feelings of vacuum, void, and loss of self; the egg made of snakes used in Druid rituals; and fertility winds capable of impregnating mares. This tradition of describing the exotic and distant was brilliantly continued by a somewhat elusive pagan author, Julius Solinus, who was nicknamed "The Teller of Many Tales," or Polyhistor. Around A.D. 250, Polyhistor published the most complete and widely-circulated collection of geographical myths, titled *Colectanea rerum memorabilium* (*Gallery of Marvelous Things*). In Solinus's world, marvels were everywhere. In Italy, one would find people who sacrificed themselves to Apollo while dancing naked over burning coals, boa constrictors that fed on cows' milk, and a lynx whose urine froze "in the hardness of a precious stone that has magnetic powers and is the color of amber."[1] In Regio, the crickets and grasshoppers still didn't dare to sing because Hercules, once disturbed by their noise, had ordered them to be silent. Further away, in Ethiopia, the dog-headed monkeys were ruled by a dog king, and one-eyed people lived along the coast. Along the Nile, one could find ants the size of wolves. In

Germany, travelers would find a creature that resembled a mule, but whose upper lip was so long that "she can't eat unless she marches backwards."[2] Elsewhere, human monstrosities included people with eight-toed feet turned backward, men with dog heads and big claws who "bark as a discourse," and people with one leg and a foot of such dimensions that they could use it like a parasol to cover themselves.[3]

A good number of these monstrosities were supposed to inhabit the opposite side of the earth. Separated from the known world by a ring of fire that circled the equator, the area was called the *antipodes*, a mythical place were all natural and divine laws were reversed. This reversal presented a curious intellectual puzzle for the early Catholic Church. How could there be an entire half of the globe inhabited by peoples who were not descendants of Adam? Even admitting the barrier was not there before the Fall, how could these people descend from Noah if, according to Scripture, the entire earth had been under water and the sole survivors landed on the top of Mount Ararat, north of the equator? Short of claiming heresy, how could the first Church fathers accept the tales of human life down South, a place where everything was supposed to be inverted, including the God-given laws of Nature? Writing about the quandary presented by this mysterious territory, Lactantius said, "One has to admit that there are men whose feet are taller than their heads, or places where things hang head down, where the trees grow downwards or the rain falls upwards," and, "What would be the wonder of the suspended gardens of Babylon if we were to believe in the suspended world of the Antipodes?"[4]

Understanding the *terra incognita* and its inhabitants became even harder from the fourth century A.D. onward as the Roman Empire crumbled. While the already well-established Christian churches were largely spared by invading barbarian tribes, the Goths and others looted and destroyed the contents of libraries and academies, erasing much of the classic knowledge from the face of Europe for the next five centuries. One of the books that disappeared during this time was Ptolemy's *Geography*, the best description at that time of the shape and topography of the known world. It was a masterpiece of geometric reasoning that had allowed cartographers to precisely project the round surface of the globe onto the flat surface of the map. With its disappearance, crucial parameters such as latitude and longitude temporarily vanished from European knowledge as cartographers kept drawing maps that now functioned more as articles of faith than as real geographic documents.

Some 600 of these medieval, faith-based maps still exist, dispersed in libraries worldwide. Even though many may have been lost, this number tells us that both artisans and their patrons were fascinated by the idea of reproducing the whole earth in one drawing. The small variations between them also tell us of a Europe dominated by Christian beliefs that were the driving force behind much of the art and scholarship of the time.

In these so-called T-O maps, the earth is presented as a circle surrounded by the "Ocean Sea." Within the circle (O), two large bodies of water form a T and separate three masses of land, each one inhabited by the descendants of one of Noah's three sons. The top of the T is the Nile–Danube, the lower portion is the Mediterranean. East is always on top, defining the orientation of the map. Above the Nile–Danube is Asia, which is inhabited by the posterity of Shem. To the north of the Mediterranean stands Europe, inhabited by the posterity of Japheth. To the south is Africa, inhabited by Ham's descendents. The center of the circle is marked by Jerusalem, the *umbilicus terrae*, or the navel of the world. The garden of Eden is always placed to the extreme east, sometimes fenced by walls of fire and abysses full of monsters, sometimes sitting on top of a mountain so high that it touches the orbit of the moon (and thus escapes the waters of the deluge).

Tales of brave people who sought the road to Paradise abound, and systematically include lavish descriptions of the

liber xiiii

europa & affrica

De. Asia & eius par

giones· quarū breuiter nomina et situs expediar
a paradiso ¶Paradisus est locus in orientis par
tus· cuius vocabulum ex greco in latinum verti
hebraice eden dicitur· quod in nostra lingua del
quod vtrumqʒ iunctum facit ortum deliciarum
genere ligni & pomiferarum arborum consitu
lignum vite. Non ibi frigus· non estus· sed per
peries· e cuius medio fons prorumpens· totum r
uiditurqʒ in quatuor nascentia flumina. Cuius

A seventh-century T-O map from Saint Isidore of Seville's Etymologiae. (Courtesy of the Rare Books Division of the Library of Congress.)

dog-headed men, pygmies, serpents, giants, talking birds, and other strange creatures found on the way through "the deep forests of India."[5] Some tales describe travelers taking to the ocean to find Paradise on some as-of-yet unknown island. Occasionally, they discovered places of such incredible beauty that it seemed as if God had conceived them for the eternal delight of the faithful. The most famous of these was the island of Saint Brendan the Navigator, first described by an Irish monk of the same name in the sixth century. Many of the stories about this mythical place picture the island as Eden. But it's always a strange, slightly perturbing Eden, with an abundance of the fabulous, such as the barnacle tree that grows birds instead of fruit, or an enormous creature with a breath so sweet that it could charm its prey and lure other animals into its den. As is true of many stories in the genre, the early medieval visions of this Eden were highly ambivalent. Later authors sometimes take this island for Ireland, and here Eden becomes quite close to hell, with inhabitants speaking in tongues and greatly enjoying incest. The island of Saint Brendan kept appearing in maps, located in different oceans, up until the seventeenth century. No wonder, then, that the Huguenots who left Europe to build their own Utopia on the island of Réunion (see Chapter 5) were convinced that their destination was the true Paradise on Earth.

Needless to say, all these "explorations," whether actual events or pieces of European folklore, kept filling the maps of the world with more and more strange creatures. Many of these were imported from oral tradition based on classic books, such as the aforementioned *Natural History* and *Gallery of Marvelous Things*. Others were free-flying interpretations of sacred texts. During the late Middle Ages, both monastic and secular writers began to convert the monsters of mythology into Christian symbols. Passages such as "Pliny told us about these things; but he spoke of marvels, and I speak of moral" abound in those studies.[6] During the fourteenth and fifteenth centuries, authors used popular poems from the preceding centuries to explain other terrible mythical beings. The horrible serpent Iaco symbolized "cholera and mental furor," while a poisonous plant in Sardonia that caused its victims to die of laughter demonstrated that "the joys of this world lead to death."7 Some special animals—such as the Indian manticore—were so horrifying and scary that the authors preferred to present them in all their terrifying glory, leaving out any moral meaning:

> It has a triple layer of teeth, the face of a man, and green eyes; it is red as blood in color, it has the body of a lion, a pointed tail with a sting like the scorpion's, and its voice is a hiss. It delights in eating human flesh. Its feet are very powerful and it can jump so well that not even the greatest well or obstacle can stop it.8

Real animals that populated these distant and marvelous lands were also given great significance. The panther, which has had religious connotations since the twelfth century, was celebrated for the alleged sweetness of its breath: "When the other animals hear its voice they gather from far and near, and follow it wherever it goes. In the same manner Our Lord Jesus Christ, the real panther, descended from heaven to save us from the Devil."9 Writing in *On the Properties of Things* in 1240, the Dominican Bartholomaeus Anglicus explained that his work will show "the enigmas from Scripture, which [are] transmitted and disguised by the Holy Spirit in the symbols and figures of properties of all things natural and artificial."10

Other more common creatures with varying moral connotations were also included in these medieval natural histories. Among the frequently mentioned animals are dogs, whose fidelity is listed by several authors in the inventory of the "visible miracles of God, that manifest themselves against the most common laws of Nature."11 The elephant, the Phoenix from Arabia, the Satyr of Ethiopia, and Barnacle Geese, the latter of which sprung from a tree covered with budding birds in the manner of flowers, were also mentioned. Following the

Augustinian tradition, authors from this period insisted that these marvels were meant to make the faithful admire creation, and, through creation, the wisdom and power of the Creator. This posture is well-illustrated in the following passage about marine monsters by the French monk Thomas de Cantipré in the thirteenth century:

> They were given to us by the Omnipotent God for the wonder of the Globe. Because in this sense they seem very marvelous, since they rarely offer themselves to the eyes of men. In truth it can be said, that God almost did not act so marvelously in many other things under the sky, except in human nature, where we can see the imprint of the Trinity. Because what under the sky can seem more marvelous than a whale?[12]

This world of fantasy should have been stripped of a good share of its monsters and marvels, at God's service and otherwise, by actual descriptions from travelers. Both the Franciscan missionaries, whose extensive thirteenth-century voyages allowed them to amass a great deal of knowledge about the lands beyond the Volga,[13] and the Crusaders had the opportunity to experience faraway lands firsthand. Interestingly, although the knowledge amassed at this time was reported to the Pope, embellished tales of their voyages with greatly exaggerated perils—not to mention the many miracles they had witnessed—spread much faster than the actual manuscripts. Thus, the world was filled with more and more layers of monstrosity awaiting mankind at every corner. The exuberance of these beliefs is well-attested by the impressive *Letter of the Prester John of the Indies*. This document, written by an anonymous author, was circulated and translated through all of Europe from the eleventh to the fifteenth century. Prester John was supposedly a Christian king living somewhere in the mysterious East. (Although he is presented as "of the Indies," his capital tends to be located in Ethiopia.) Wherever he may have been, he claimed to have unbelievable wealth and ever-growing numbers of bizarre armies made of

strange creatures with curious powers, all of which he offered
to the Pope to help save the world from the Muslim menace.
Among other things, Prester John claimed to have flocks of
birds that could drag a camel through the air, armies of pygmies, soldiers that were half man and half dog, rivers paved
with gold and silver springing from jewel-filled subterranean
mines, and an entire province populated solely by female can-
nibals who were fearsome warriors. With each translation or
new copy, the numbers and marvels of his warring menagerie
grew to increasingly impressive proportions, giving us a clear
image of natural history in the European medieval mind.

Did these dreams come to an end with the onset of the
European voyages and discoveries? By the end of the eighteenth century, Solinus's creatures had been erased from the
maps, the island of Saint Brendan was no more, and the
famous dog-headed people (known to medieval scholars as
"cinocephali") were nowhere to be found. Did Europe accept
this loss of mystery and magic without a fight?

Of course not.

First of all, as eight-toed humans left the scene, real societies came into view. While the explorers often found these
societies just as curious as their mythical counterparts, many
of them also saw the well-being, happiness, and tranquility
that had long been lost in Europe. Travel literature from the
sixteenth century onward, often written by churchmen,
praised the harmonious ways of the newly discovered societies. According to these authors, American Indians lived in
peace and perfect equilibrium with nature, wise Chinese
rulers dispensed generous and friendly laws to ensure the
happiness of their subjects, and quiet Africans found cures to
all evils using their god-given knowledge about the magic
properties of their environment.

This literary trend soon evolved from report to satire, as
brilliantly demonstrated by Jonathan Swift's *Gulliver's
Travels*. Placing the evils of Europe in a faraway land was a
safe way to criticize governments without incurring too much

Over the years, many have come to believe that the phrase "Here be dragons" was always placed on a map where the mapmaker's knowledge ended. While the phrase does articulate the fears that many medieval cartographers had about distant lands, the phrase only appeared on this sixteenth-century copper Lenox Globe. (Rare Books Division, The New York Public Library, Astor, Lenox, and Tilden Foundation.)

risk of retaliation, just as depicting perfect societies in fictional islands was a safe way of engaging in popular propaganda. This literary movement coincided with Jean-Jacques Rousseau's landmark description of the "noble savage," based on the then-popular notion that *Là-bas, on était bien* (Out there, we were well). For French thinker Michel Eyquem de Montaigne and his contemporaries, this *là-bas* was America, with peaceful Indians and vast rich spaces open to Europe as a grand opportunity to start anew and reinvent happiness.

As America became better known, more colonized, and thus inadequate for such dreams, philosophers and satirists had to find yet another place on the map for their perfect societies. By the time Denis Diderot began writing in the eighteenth century, the idealized place had become Tahiti or some other island in the South Pacific, depending on the author. Within the satiric vein, Louis Froquet published his provocative *Terre Australe Connue*, telling readers of a Polynesian island inhabited by naked hermaphrodites living with no rules, no hierarchies, and no god. For the first time, the religious authorities protested.

The idea of this new, marvelous place was pretty much outmoded by the nineteenth century, but meanwhile it had inflamed many spirits to go look for the perfect place for their own particular utopia—which, as we shall see, was a crucial factor in the discovery of the dodo bird.

As they traveled, the European discoverers were certainly not finding manticores or iagos in their travels, but they were encountering a brand-new bestiary, in all ways as bizarre and fascinating as the tales from the old maps. New and exotic natural objects of all sorts, coming from all quadrants of the compass, soon became mandatory ornaments in the treasures of the rich and powerful. These ornaments could be stones or shells painstakingly decorated with gold, silver, enamel, and jewels by the patron's artisans; or they could be displayed simply as they had been found. In either case, there was absolutely no repugnance in enhancing their value with symbolic or magical connotations: ostrich eggs decorated churches as eggs of the griffin, sea-lion canines were raised over the altars as unicorns' horns, and embalmed crocodiles were the rage of rich convents.

It is arguable, though, that nothing had more power over the abbot's flocks or over the emperor's subjects than the live specimens brought from afar that miraculously managed to survive the long road home. Nothing made a powerful man of the sixteenth century seem more powerful than a copious menagerie of exotic beasts. Likewise, nothing made a street merchant enjoy such immediate financial profit as the display—inside a tent so that the passerby had to let go of some coins to see the thing with their own eyes—of a never-before-seen animal alive on stage. And, for both purposes, the stranger the animal the better.

That's why the dodo bird of Mauritius was shipped to European ports and traded between aristocrats and street vendors.

The rest is history.

1 C. Julius Solinus and Arthur Golding (trans.), *The Worthie Work of Iulius Solinus Polyhistor: Contayning Many Noble Actions of Humaine Creatures, with the Secretes of Nature in Beastes, Fyshes, Foules, and Serpents: Trees, Plants, and the Vertue of Precious Stones: With Diuers Countryes, Citties and People: Verie Pleasant and Full of Recreation for All Sorts of People*, London: Printed at I. Charlewoode for Thomas Hacket, 1587.
2 *Ibid.*
3 *Ibid.*
4 Lactantius and Michel Perrin (ed., trans.), *L'ouvrage du Dieu Createur*, Paris: Editions du Cerf, 1974. Translated from French by Clara Pinto-Correia.

5 Arthur Percival Newton (ed.), *Travel and Travellers of the Middle Ages*, London: K. Paul, Trench, Trubner & Co., Ltd.; New York: A. A. Knopf, 1926.

6 Lorraine Daston and Katharine Park, *Wonders and the Order of Nature, 1150–1750*, New York: Zone Books, 1998.

7 *Ibid.*

8 *Ibid.*

9 *Ibid.*

10 Bartholomaeus Anglicus, "*De proprietatibus rerum*," in: M. C. Seymour *et al.* (eds.), *On the Properties of Things: John Trevisa's Translation of Bartholomaeus Anglicus De proprietatibus rerum: A Critical Text*, Oxford: Clarendon Press, 1975–88.

11 Lorraine Daston and Katharine Park, *Wonders and the Order of Nature, 1150–1750*. New York: Zone Books, 1998.

12 *Ibid.*

13 William of Robruck, leader of the second Franciscan expedition to Tartary, for instance, noted for the first time that the Chinese "drawings" were, in fact, a different alphabet. He also accurately described the rites of Buddhist monks.

Chapter 2

The Discovery

To the south of the equator, off the southeast coast of Africa, a number of island groups pierce the waters of the Indian Ocean. Close to the African mainland are the Tanzanian islands Pemba, Zanzibar, and Mafia. Farther offshore, to the southeast, are the Comoros Islands, followed by the northern tip of the great island of Madagascar. Northeast of Madagascar, we approach the Seychelles, then bear east to the Chagos Archipelago, halfway across the ocean to Indonesia. But let's swing back southwest to the scene of our tale.

Between Madagascar and Australia, the Indian Ocean stretches for thousands of miles, virtually unbroken by land. The only solid land in the open sea is a group of three islands strung along the twentieth parallel of southern latitude.

These are the Mascarenes: Mauritius (500 miles east of Madagascar), Réunion (southwest of Mauritius), and Rodrigues (smallest of the three and farther to the east). Réunion and Mauritius are only 100 miles apart. Rodrigues (occasionally spelled "Rodriguez") is the farthest east, 360 miles from Mauritius and 450 from Réunion. Mauritius is an independent republic. It has an area of 720 square miles and a population of more than 1 million, making it one of the most densely populated countries on earth. Réunion, with an area of 970 square miles and a population of some 670,000, is an overseas *département* of France, with representatives in the French parliament. Rodrigues, today a part of Mauritius, has an area of just 40 square miles and a population of 37,000.

The Mascarene Islands were produced by volcanic activity, but they are not of the same age. Mauritius is by far the oldest and has no volcanic activity, whereas Réunion, the largest of the three, still has an active volcano.

Some geologists have suggested that all these islands were once part of the ancient super-continent of Gondwanaland. However, the distribution of plant and animal life on the islands calls this link into question, since each island has its own unique flora and fauna. The sharp segregation of life-forms in the Mascarenes can be explained at least in part by the fact that the three islands existed for millions of years undisturbed by human visitors who might have carried plants or animals from one island to another.

There is no record that the islands were ever settled or even extensively explored by human beings before the sixteenth century. Even the Malaysian emigrants who crossed the Indian Ocean in some ancient period to become the ancestors of the Merinas, the rulers of Madagascar from the late sixteenth century until the nineteenth century, appear to have left no trace in the Mascarenes.

It is possible that at the time of King Solomon (tenth century B.C.), Phoenician expeditions sailing from Elath on the Gulf of Aqaba in the Red Sea penetrated as far south in the Indian Ocean as the shores of Mozambique. Around the same time, the Comoros Islands were visited by Arabs or Jews of the biblical land of Edom sailing from the Red Sea. Still, Greek sources suggest that only the northern half of the Indian Ocean was known in the classical period. It was described in an anonymous Greek book of sailing directions compiled between the first and third centuries A.D. The book calls these waters the Erythraean Sea or Red Sea. (What we now call the Red Sea was then called the Arabian Gulf.) It includes a reliable description of the African east coast, which

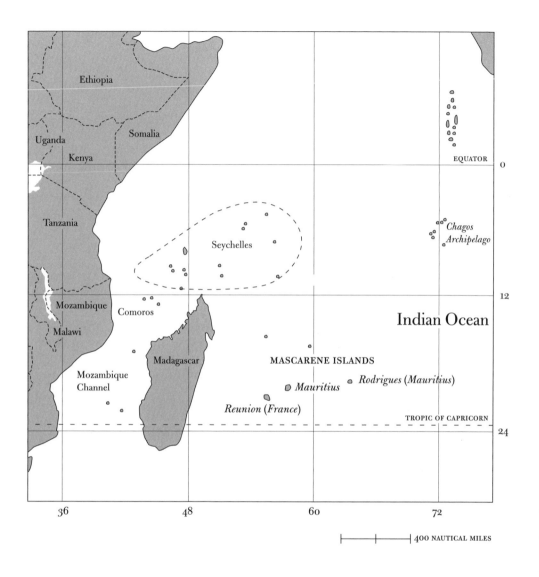

The Mascarene Islands. (Based on a map provided by The
General Libraries, The University of Texas at Austin.)

was known to contemporary seamen as Azania. However, Madagascar and nearby archipelagos are never mentioned.

Ptolemy described the coast of Azania again in the second century A.D., but added nothing new. Four centuries later, the work of Cosmas Indicopleustes—an Egyptian monk and one of a small number of writers in the early Middle Ages who claimed that the earth was flat—added little to the picture, for Azania remained outside the main trade routes of the Indian Ocean. The small vessels that plied this trade were designed to sail before the monsoon winds; thus, they were so light that they dispensed with nails and were simply sewn together. They were sufficient to hug the African coast, the shores of the Red Sea and the Persian Gulf, and the Malabar Coast of India, but they were unable to face the open sea to the south.

By the end of the first millennium, or perhaps even earlier, the Mascarenes were visited by Arab merchants. Their ships roved freely and boldly in the Indian Ocean, from the Near East to Africa, India, and even beyond to China from the tenth century through the twelfth and thirteenth centuries. The three islands are shown on at least one old Arab map, but those visitors made no settlements, as they did in the Comoros, and their short visits left no historical traces and had no impact upon local flora and fauna.

By the ninth century, Azania appears in Arabian maps under the name of Zanj, bathed by the Sea of Zanj. The Arab voyagers who settled on the African east coast intermarried with the local Zanj people, giving rise to the Islamic civilization of Swahili, a term that literally means "coastal." This culture spread via a number of loosely linked trading posts, never penetrating far inland, and reached its peak in the twelfth century. It was propelled by the introduction of a new type of ship, the dhow—misleadingly called Arabian, since *dhow*, or rather *daw*, is a Swahili word. (The Arabs preferred the term *sambouk* to refer to any kind of dhow.) The new dhows still had their planks sewn together, but they were strong enough for the Swahili people to cross the sea to Madagascar and its neighboring islands. The Mascarenes were among those explored by the Swahili in this period. Bearing Arab names, they appear on the world map of Cantino, published in 1502. Mauritius is *Dina Mozare*, Réunion is *Dina Margabim*, and Rodrigues is *Dina Arobi*, the desert island.

———

Islam swirled by the Mascarenes, but it had no interest in laying claim to empty land. Christendom would play that role.

There were strong incentives for establishing a European Christian presence in the Indian Ocean, where a Muslim trading empire controlled a large part of the huge commerce in spices and textiles that existed between Asia and the Christian lands of Europe. The first Europeans to embark on the dangerous adventure through uncharted waters to seek a sea route to Asia were the Portuguese, who operated in the spirit of a religious as well as commercial crusade.

What was the trigger for the Portuguese démarche? More than anything it was the vision of one man—Prince Henry of Portugal (1394–1460), son of the Portuguese king John I of Aviz, who became known to history as Henry the Navigator. According to the historian João de Barros (1496–1570), writing in his voluminous *Décadas* a century later, by the mid-1400s Prince Henry had little to do because "in his kingdom there were no more Moors to conquer; since the kings his grandparents had thrown them all overseas and onto the parts of Africa."[1] Of course, Henry could have carried the fight to Morocco and, in the process, acquired rich lands for the Portuguese crown. But he would have had to do so as a "sent captain, and not as a conqueror, since the conqueror had to be the king himself." Because his older brother, Duarte, was next in line for the thrown, thus denying Henry the right to be the "Conquerer" of North Africa, the prince set his sights elsewhere.

After fighting the Moors in "distant and unconquered parts of Spain," Henry turned his attention to what he believed to be a vast realm on the west coast of Africa.[2] At the port of Lagos, not far from Sagres on Cape St. Vincent, Henry established a school for mariners and began sending expeditions south for trading, preying on local shipping, and exploration, decades before the Spanish crown lent an ear to the schemes of one Christopher Columbus.

In 1433, one of Henry's captains, Gil Eanes, was the first to brave the crossing of the dangerous Cape Bojador on the northwest African coast, south of the Canary Islands. Shortly thereafter, Portugal triumphed by seizing, as Barros notes, "the oil and leather of the million wolves of the sea [seals] they could kill in those places."[3]

Along the way, Henry's mariners claimed for the prince "some more tongue of land we can find."[4] Among the lands seized were Madeira and Porto Santo, the Azores, Cape Verde, Guinea, and São Tomé and Príncipe. Then, the Portuguese began taking possession of nearly all the land along the African coast that was approachable by boat, beginning with the area now occupied by the Ivory Coast and Sierra Leone. After these lands had been seized, the Portuguese claimed what is now Angola. Next, the Portuguese were off to the Cape.

After Henry's death, when King Manuel ascended the Portuguese throne,

> he also inherited the continuation of the enterprise his predecessors had started, that being the discovery of the Orient through this ocean sea of ours, that so much industry, and work, and expenses had cost for seventy five years: immediately in his first year as king he wanted to show his will to persevere.[5]

The perseverance paid off. In 1494, the Portuguese explorer Bartolomeu Dias sailed around the Cape of Torments, since known as the Cape of Good Hope, and generally referred to as "the Cape" because of its importance in east–west navigation. In 1497, the fleet of Vasco da Gama,

PRINCE HENRY
OF
PORTUGALL

CEUTA

FAC-SIMILE OF AN ANTIQUE AND EXCEEDINGLY RARE ENGRAVING MADE IN HOLLAND ABOUT THREE HUNDRED YEARS AGO.

Prince Henry the Navigator (1394–1463)
third son of John I

In addition to personally organizing and funding many sailing expeditions, Henry the Navigator employed a number of cartographers who created sophisticated maps, allowing sailors to accumulate and share information. The resulting knowledge was gathered in what was called the "School of Sagres." The name was a tribute to the town of Sagres (located on Cape St. Vincent, the southwestern tip of Portugal) near the port of Lagos, from where the caravels left on their sailing voyages. This school—simultaneously intellectual, technological and commercial—was the first organized project directly related to the European discoveries. (Picture Collection, The Branch Libraries, New York Public Library.)

"nobleman of the King's house," followed Dias's trail and entered the Indian Ocean.[6] An Arab pilot showed da Gama the way to Goa, a port on the west coast of India, where he arrived in 1498. In the next few years other Portuguese expeditions followed. In 1494, the Treaty of Tordesillas divided the world along the Atlantic Ocean, the West belonging to Spain and the East belonging to Portugal, to put an end to endless bouts of petty disputes over such small possessions as the Canary Islands and parts of the shore of Guinea. With the Spanish distracted by the exploration of America as an alternate route to the East Indies, then as a source of wealth unto itself, Portugal was the only European power operating in the Indian Ocean for several decades.

In February 1507, Captain Diogo Fernandes Pereira sailed eastward from Madagascar in an expedition set for Malinda, a seaport near Goa, under the command of the powerful Portuguese Viceroy of India, Afonso de Albuquerque. Pereira came upon the island of Réunion and named it Santa Apolonia, after the Portuguese patron saint of February 9, the day of his discovery. After replenishing the supplies of drinking water and food, Pereira sailed on, reaching the island of Mauritius, which he named Ilha do Cerne—Swan Island—after his ship, *Cerne* (Swan), the first to enter a bay in Mauritius. The first visitors set sail again, still seeking the coveted East without even bothering to claim Mauritius for their crown. On the way, they found the third island in the group, Rodrigues, which, for reasons still unclear, Pereira named Domingo Frias. Having landed on the islands only to restock fresh water and fruit, Pereira's men saw nothing unusual. More importantly, they had no impact on the ecology of the islands.

In 1513, Mauritius (Ilha do Cerne) was approached by another Portuguese skipper, the nobleman Pedro (or Pero, in

Pedro Mascarenhas, who gave the name Mascarenes to the group of islands now known as Mauritius, Réunion, and Rodrigues. (The Royal Geographical Society, London.)

archaic Portuguese) Mascarenhas, nephew of a captain of horsemen for kings João II and Manuel, whose good services had been rewarded with a title of nobility. Pedro Mascarenhas dropped anchor in Ilha do Cerne and promptly renamed it Mascarenhas, after himself. He used the same self-reference to call the entire archipelago the Mascarenhas (later transliterated to Mascarenes, which would stick even if the Portuguese names for the two larger islands would not). However, conditioned by the design of their colonial interests, particularly their sailing route to India and their commerce in the Far East, the Portuguese didn't pay much attention to this discovery.

If anything, the sailors sent ashore noticed a dumb-looking, fat, ugly, flightless bird. Although they wrote little about this creature in their official voyage accounts, we know that they saw it because of the name it has borne ever since. *Doudo* means "crazy" in sixteenth-century Portuguese.

And what a strange bird it was—a weird, mammoth turkey with the face of a warrior pigeon, an ugly joke of nature. It was a slow, heavy, clumsy creature that was seemingly silent and couldn't even fly, and which appeared unable to run away from men. A crazy bird. Or, in Portuguese, a *pássaro doido*, or rather, in the way Portuguese was spoken and written in the sixteenth century, a *pássaro doudo*.

Round as a sack, it had an immense head, and its black bill ended in a great snubbed hook. The plumage was ash-gray, its breast and tail whitish, and its wings yellowish white. Unable to fly, it waddled along on short yellow legs and big splay-toed feet. When it tried to run, it jogged so clumsily, according to sailors' narratives, that its plump belly scraped the ground. It was a placid, slow-moving beast, to be described by seventeenth-century chroniclers as "loyal to its mate and dedicated to its chicks."[7]

Being flightless meant that the dodo had no way of climbing into the trees, so it made its nest on the ground, presumably deep in the woods. Here, during each mating season, the

The dodo of Mauritius. From the descriptions of the European sailors who first encountered the beast, it was huge, fat, and easy to carry, and could feed many people at once with food still left over. (Illustration by Jacques Hnizdovsky for Robert Silverberg, *The Auk, the Dodo, and the Oryx: Vanished and Vanishing Creatures*, New York: Thomas Y. Crowell Company, 1967.)

females laid only one egg, building no defenses around it. Those had never been necessary, because the dodo had never encountered any enemies until the Europeans discovered the islands. For the same reason, the dodo had no idea what fear might be, thus acting "dumb" or "feeble of wit," as so many mariners would describe it.

———————————

It was after the explorers' departure that the tragedy began to unfold. The bird existed only on Mauritius, and nowhere else on the planet. This was no accident, for islands often develop their own unique inhabitants. Since they are isolated from the rest of the world for perhaps millions of years, the dots of land surrounded by sea allow their flora and fauna to evolve and develop according to their own rules. Such island species progressively become better equipped to make the maximum use of the resources in their self-contained environment. And, over time, they lose much of their resemblance to their closest relatives living elsewhere. Small, out of the way islands are the places where Nature plays her most bizarre games, because they're almost entirely sheltered from the outside world and thus spared from the much more complex laws that govern life's mosaic in the broad expanses of continental land.

The intense drama, one that was relentlessly staged and re-staged during the era of European discovery, occurs as soon as islands are invaded by alien visitors. The island animals and plants, accustomed to being left alone in their long-established and unperturbed harmony, often find themselves powerless before invaders. Taken by surprise, they might shake and struggle, but they find it hard to put up a strong resistance. What would they know about putting up strong resistance, anyway? The animals are lacking in fear and the instinct for self-defense because they never needed them. Therefore, the newcomers take their place effortlessly, and the natives slowly slide further and further into the shade, in deeper and deeper silence, until the shade gives way to oblivion.

The explorers had stayed on Mauritius long enough to leave in their wake a first tidal wave of invaders, setting in motion the now-familiar process that can destroy native wildlife. As the boats disappeared in the distance below the rounded curve of the horizon, a curious swarm of restless aliens was now exploring the island's grounds. The Portuguese had just introduced the first monkeys and goats to the island. This practice was actually quite common. Before departing, sailors would leave behind animals they liked to eat so that the animals would reproduce, and the next people stopping by the island would have good meat to eat.

When the ship left Mauritius, it also left a herd of unanticipated colonizers—hundreds of rats escaping in a flurry from their dark and moldy refuge in the bowels of the vessel.

The Portuguese had no particular commercial or strategic interest in the small archipelago named after one of their own. In 1510, the Portuguese captured Goa on the Malabar Coast of India, and in 1511 seized Malacca, a port on the Malaysia peninsula that commanded a key passage in the trade route to the Spice Islands and China beyond. In Asia, their principal commercial bases were at Goa; on the island of Sri Lanka, then called Taprobana and later Ceilão (whence the English "Ceylon"); at Malaca; and in Timor, in the Indonesian chain. (Timor would remain a Portuguese colony until 1975.) East African cities from Sophala, in modern Mozambique, to Mogadishu, near the horn of Africa in modern Somalia, were taken over, some by force.

In the west Indian Ocean, in the first decade of the sixteenth century, the Portuguese had turned their attention to Madagascar (São Lourenco), the Comoros (Ilhas do Comoro), and the Seychelles (Sete Irmanas). In the next ten years, they charted Mauritius (Ilha do Cerne, Ilha do Cirne, or Ilha do Cisne, all more or less contemporary terms for "Swan Island"), and Réunion (Santa Apolonia). Rodrigues (first Domingo Frias, then Diogo Rodriguez, and sometimes even Diego Ruiz) and the Chagos (Chagas) Archipelago would appear in Portuguese maps in 1538. The early sixteenth-century exploration of the area by the Portuguese resulted in the first European scientific charts of the Indian Ocean, coinciding with the dominating presence of Portugal in a vast region taking in Southeast Asia, the entire coast and the islands of the Indian Ocean, and both the east and west coasts of Africa. This presence is acknowledged by the persistence of Portuguese names in the region to this day.

———

Although Portuguese vessels of the sixteenth century were much more sophisticated than the Arabian dhows, and their nautical instruments were much further advanced, the ships on the *Carreira das Índias* (the "Indian line," also known later as the "inner route") still made the passage from Africa to the east with the aid of the seasonal monsoons, as their predecessors had always done. The term *Carreira das Índias* applied specifically to the round-trip voyage made by Portuguese Indiamen between Lisbon and the great port of Goa in the days of sail. The seasonal winds of the tropics were the determining factor, and under the most favorable conditions the voyage, including the stopover at Goa, took about a year and a half.

  The southwest monsoon, which normally begins on the west coast of India in early June, had the effect of virtually closing all harbors in the region from the end of May to the beginning of September. Thus, the trading season lasted from September to April. Portuguese ships, overcrowded with an assortment of passengers ranging from soldiers and missionaries to men and women seeking their fortunes abroad, sought to leave Lisbon before Easter. If all went well, they could round the Cape of Good Hope in time to catch the tail end of the southwest monsoon winds off the east African coast north of the equator, which would bring them to Goa in September or October. On the return leg, the ships overloaded with spices, precious china, and other exotic goods left Goa with the northeast monsoon around Christmas, so as to round the Cape before the region was struck by the storms that started there in May.

The ships engaged in these long voyages were principally a kind of galleon then called nau, or great ship: a merchant-type vessel evolved from those used by the Venetians and Genovese in the Middle Ages, and developed to spectacular sizes by the Portuguese during the trading fever of the sixteenth and seventeenth centuries. The *Nau da Carreira das Índias* was broad in the beam, with three or more flush decks, a high poop, and a forecastle, lightly gunned for her size and often a sluggish sailer. The early naus weighed about 400 tons loaded, but eventually weighed over 2000 tons, becoming the largest vessels afloat at the turn of the sixteenth century, rivaled only by the greatest of Spanish Manila galleons.

Some of the best and biggest naus were built in Portuguese India. The Portuguese had quickly recognized the superiority of Indian teak for ships over European pine and even oak. Cochin, near the subcontinent's southern tip, Bassein, in present-day Myanmar, and to a lesser extent Daman, north of Bombay, all became important shipbuilding centers. Ships for Portugal were built on contract with the local raja in the case of Cochin. The great royal arsenal and dockyard at Goa was probably the most highly organized Indian enterprise at the time of the Mogul Empire. A royal order of 1585, repeated nine years later, emphasized the importance of building the *Naus da Carreira* in India rather than in Portugal "both because experience as has shown that those which are built there last much longer than those built in this kingdom, as also because they are cheaper and stronger, and because timber for these Naus is increasingly hard to get here."[8]

The Mascarenes and the Seychelles were too far away from the main trade route to have any impact on shipbuilding. Besides, the Portuguese could find all the timber they needed in Brazil and India, and therefore developed no interest in the ebony of Mauritius.

Trying to establish support bases for the *Carreira das Índias* along the African east coast, the Portuguese seized Mozambique and concentrated their efforts there. If any of the nearby archipelagos would have been of any strategic interest for this supporting role, then it would be the Comoros, not the Mascarenes. But efforts to colonize the Comoros were never quite successful. So, Mozambique retained a leading part in securing the India trade and would remain a Portuguese colony until 1975. Mauritius and its dodos were simply overlooked.

Around the same time, however, sailors calling at the other two Mascarene islands discovered the dodo's relatives. A similar bird, white in color, lived in quiet isolation on Réunion. They simply called it the *white dodo*.

And in the cliffs of the rocky protuberance called Rodrigues, a cousin with a long neck hid in the woods. Because this bird was never observed in groups, they called it the *solitaire*.

The dodo family had been registered just in time to perish, because sailors found that dodos were easy to catch and good to eat. Before the arrival of Europeans, there weren't any carnivores in the Mascarenes. Hence the birds in the dodo family had never learned to run away. Since their main food was snails and nuts, the only tools of survival they had ever needed were two solid legs and one robust hooked beak. The islands had never taught them to use anything else, and the dodo family had always fared perfectly well. They had gotten so used to their tranquil lifestyle that slowly, very

The white dodo of Réunion. The first European to spot it was a Dutch sailor named Bontekoe, who described it as a sort of gigantic goose and was particularly impressed by the number of meals these birds could provide for his crew. (Illustration by Jacques Hnizdovsky for Robert Silverberg, *The Auk, the Dodo, and the Oryx: Vanished and Vanishing Creatures*, New York: Thomas Y. Crowell Company, 1967.)

The long-necked solitaire of Rodrigues. This bird was first described by the French Huguenot François Leguat, who was a man of immense spirituality and never talked about how many men a bird this size could feed. Instead, he wrote at length about the beaty and graciousness of this Creature of God. (Illustration by Jacques Hnizdovsky for Robert Silverberg, *The Auk, the Dodo, and the Oryx: Vanished and Vanishing Creatures*, New York: Thomas Y. Crowell Company, 1967.)

slowly, as these things always happen, their wings grew shorter, while their legs grew bulkier.

Therefore, confirming the phenomenon of island speciation and extinction elsewhere, generation after generation, ever so slightly yet so steadily, the dodo of Mauritius and its cousins, the white dodo of Réunion and the solitaire of Rodrigues, gradually evolved into an easy prey for their first predators. Hunters had never been a part of the islands' microcosm, where life had never been a dangerous game in which you'd better always be on guard. When the people started invading its island, followed by all their clever, hungry beasts, all the dodo could do was to try to run. But it couldn't run very fast and had nowhere to go.

———

As time passed and the Portuguese stayed away, the ships of other nations must have passed through the Mascarenes, perhaps the English and the French, but none really took the pains to claim the islands as their own, nor did written accounts of such visits appear. Instead, it fell to the Dutch, who became the second great European player in the Indian Ocean. It was the Dutch who managed to slip into the wake of the Portuguese and establish solid settlements and claim valuable possessions before the English or the French. This feat involved a complicated give and take, for Portugal and the Netherlands had a tense relationship, especially between 1581 and 1640, when Portugal was annexed to Spain by Philip II while the two were officially at war. But, on the whole, informal exchanges of all sorts took place between the Portuguese and the Dutch, since they were sharing similar maritime fates and needed each other as much as they distrusted each other.

The career of the Dutch adventurer Jan Huygen van Linschoten (1563–1611), who left a remarkable account of his travels, illuminates the complex relationship between the Netherlands and Portugal in the sixteenth century. *The Voyage*

of John Huygen van Linschoten to the East Indies, first translated into English in 1598, provides an excellent demonstra-
tion of colonial chess at work, with players from Portugal,
Spain, the Netherlands, and England constantly moving their
pawns over the vast chessboard that was the East Indies and
all the sea routes leading in and out. Van Linschoten was born
in Haarlem, Holland, in 1563 and grew up under the occupation of his homeland by the Spanish. During much of the sixteenth century, the Netherlands was fighting for independ-
ence from Spain, but commercial relations between the two
sides were never really abandoned by either one, since the
Dutch market was then indispensable to the prosperity of the
Indian trade of both Spain and Portugal. Jan was a studious
youth who "took no small delight in the reading of histories
and strange adventures" and developed a "strong desire to see
a little of the world."9

At about 16, Jan left his parents to join his brothers, who
had a business in Seville. From there he moved on to work
with a merchant in Lisbon. During the two years of succession
wars between Portugal and Spain following the death of the
young King Sebastian of Portugal in 1578, van Linschoten
joined several military regiments and traveled in the two
countries, eventually settling in the Portuguese capital. Trade
did not prosper during the Portuguese succession war, so the
young van Linschoten decided to follow one of his brothers
and seek employment in the *Carreira das Índias*. He and his
brother procured passage on a nau in the party of João
Vicente da Fonseca, the newly appointed Archbishop of Goa.

The Dutchman would lead a very Portuguese life during
the six years he spent in Goa with the Archbishop. A short
time after his arrival, he recorded a typical incident. Two
English merchants were arrested, ostensibly for espionage on
behalf of a pretender to the throne of Portugal. The real moti-
vation for the arrest appears to have been in retaliation for
attacks carried out against the Portuguese in the Moluccas
region of eastern Indonesia, otherwise known as the Spice

Islands, by the merchants' countryman Sir Francis Drake. Drake was feared by the Portuguese, whose hold over the Spice Islands was threatened by the English admiral's promise to the inhabitants to return and "expel the oppressors."[10] Van Linschoten and a fellow Dutchman intervened on behalf of the English merchants with the Archbishop and procured their release.

While in Goa, he sought opportunities to travel to China and Japan, which never materialized, but he did travel widely in the Indies and became a close observer of both the mores of the Portuguese traders and the opportunities the Indies offered in terms of wealth.

Arriving in Java, for example, van Linschoten noted the abundance of rice ("and other things necessary to life"), cattle, sheep, fowl, garlic, "nutmegs and all sorts of spices," "other things they take to Malacca" (the Moluccas), and "very good" pepper, as well as "incense, camphor and diamonds." Then he turned to matters of lifestyle:

> Several Portuguese men in India joined local women in marriage. Their children are yellowish, although some women are quite beautiful. The mixed children are not all that different from the Portuguese, and, since the children of Portuguese parents born here also tend to be more yellowish than their forebears, by the third or fourth generation they cannot be distinguished from the natives. . . .
>
> The Portuguese, Mixed and Christian, take rich and magnificent care of their family. They commonly have ten to twenty servants, according to their means. Those who are married have their homes magnificently decorated and furnished, namely in what concerns lingerie. They always dress finely to leave the house, and are of extreme courtesy with each other, using repeated salutations and bendings of the loins and kissing hands, and even more so when they go to the mass, where their servants prepare their seats for them. . . .[11]

Van Linschoten next focuses on women:

They take the meat with their hands, considering the use of
the spoon ridiculous and uncivilized. They drink from bottles
shaped like vases with a devise called *gorgoleta* that allows
them to swallow the licquour without touching the bottle,
and find this use quite gentle. Those who are not used to this
way of drinking can't first handle such bottles without spilling
licquour all over their chest. In fact those who've just arrived,
and don't know the local costumes, and are not yet grave
enough as they walk, can't cross the street without being
mocked. But they learn the new uses fast and take to them
with pleasure.[12]

This curious vignette of colonial daily life is followed by
steamy details of how the East tends to corrupt European
families:

The jealousy of husbands for their wives and daughters is
very big, and they don't let anyone see them, except when the
families spend time with their children and some close
friends in some orchards, always with a group of slaves and
soldiers standing nearby to guard them. As soon as their boys
[in the Portuguese families] turn fifteen, they are moved to
their own quarters, away from their mothers and sisters, this
being due to the enormous and strange lascivity of these
women, that triggers quite constant acts of incest that lead to
the children being killed when discovered in the act, or to the
raging husbands killing their wives in their blind ardor. There
are few married women here who keep their marital chastity,
& don't have a ruffian among soldiers, merchants, or
chamber-boys for amusements no one can stop them from
having, since they often use herbs for this purpose. There is
even a juice called *Dutroa* that they give to their husbands &
plunges them in a stupor or deepest sleep as though they had
totally lost their senses, and any one would think them dead,
during which time women may do as they please, as this time
is as long as 24 hours. The only means to wake up the
husbands from these trances is to wash their feet with cold
water, but even then they can't remember anything that
happened in their sleep.[13]

When his archbishop patron died, van Linschoten decided to return to Europe, which took three years, including a
two-year stay in the Azores, during which he made careful

note of British attempts to raid Spanish and Portuguese ships returning from the Indies.

In 1592, van Linschoten finally reached Lisbon, then sailed to the Netherlands, returning home after an absence of nearly 13 years. Four years after his return, van Linschoten's narrative of his adventures was published in the Netherlands. In the meantime, he signed on to a Dutch expedition organized with the approval of Maurice of Nassau, Prince of Orange and the Netherlands' chief magistrate and ruler, to explore the possibility of a northeast route to the Pacific and thence to the Indies.

Three ships sailed north from the Dutch port of Texel in June 1594, up the coast of Norway, then turned east into the Arctic Ocean. But they returned to the Netherlands in September of the same year. They had reached the Kara Sea, and found what appeared to be a passage to the east but had been forced to turn back because of ice.

Van Linschoten reported to Maurice of Nassau his belief that a northern route to China and India had now been discovered, and he succeeded in imparting his hopes to many of his compatriots. The following year, seven ships were fitted out to continue the exploration, but this time the ships didn't even reach the Kara Sea because of the ice. The Dutch government resolved to make no further attempt at the public expense. A third attempt, privately financed, failed miserably.

Meanwhile, early publication of portions of van Linschoten's adventures had prompted Maurice of Nassau to send a Dutch fleet into the Indian Ocean under Captain Cornelis Houtman to follow the route of the Portuguese. Maurice stipulated that the Dutch ships should as much as possible avoid conflicts with the Portuguese and seek friendly relations with the inhabitants of the lands they visited. Van Linschoten had already pointed out the importance of the trade with Java, remarking that on this island "men might well traffic without any hindrance, for that the Portingales come not thither because great numbers of Java come themselves unto Malacca

MAURICE PRINCE D'ORANGE.
Comte de Naſsan Meurs &c Gouverneur Ge
neral deſ Provinces-Unies et Chevalier de
l'Ordre de la Jarreſiere.

Maurice of Nassau, captain general and admiral of the United Netherlands, to whom Mauritius owes its name. A proud member of the House of Orange, Maurice was lenient regarding religious matters, making the Netherlands a safe haven for spiritual refugees from adjacent sections of Europe. In return, these refugees made important contributions to the Netherlands' cultural and commercial sectors, culminating with a bout of maritime expansion so robust it even threatened the Portuguese possession of Brazil (and Brazilian gold). (Picture Collection, The Branch Libraries, New York Public Library.)

to sell their wares."[14] It so happened Java was the first land in the Indies that the Dutch reached.

But van Linschoten erred in assuming that the Portuguese would not go to Java. When the Dutch arrived at the city of Bantam, they found that Portuguese merchants already operated there. This preliminary disappointment notwithstanding, the success of subsequent Dutch voyages proved van Linschoten right in directing the Dutch attention to Java. The choice of this island for their headquarters was a main cause for the rapid rise of Dutch power over the Indies.

Moreover, the full publication of van Linschoten's narrative was pivotal in revealing to the Netherlands that the colonial empire of the Portuguese in the East was starting to rot away, and an energetic rival had a good chance of supplanting it. The measure of van Linschoten's influence can be found in the publishing history of his book. English and German translations were out in 1598; two Latin translations were published in 1599, one at Frankfurt and one at Amsterdam; a French translation followed in the same year.

In 1598, Maurice of Nassau sent Admiral Jacob Cornelius van Neck on a voyage of discovery with a fleet of eight ships. It was this expedition that sighted the Mascarenes and dropped anchor at Ilha do Cerne. Van Neck renamed the island Mauritius after Maurice; this is the name that would stick.

From this point on, the Netherlands was beset with navigation fever. The East India Company of Amsterdam fitted out eight ships that sailed south in early February 1599, followed by three more departing in May. On July 8, the first three of van Neck's eight ships returned. They were quickly unloaded and given orders to set sail again. At the same time, another group of Amsterdam merchants formed a new company that sent out four ships in December 1599, accompanied by four more from the old company. These eight ships

Admiral Jacob Cornelius van Neck, commander of the second Dutch voyage that followed a route encompassing Mauritius, declared the island a Dutch possession and was the first European to fully explore it. He is often credited as being the first to draw the dodo. The journal describing his travels was enormously successful and had several translations soon after the first printing. (Courtesy of Rijksmuseum Amsterdam.)

returned two years later, loaded with riches. But, before their return, the new company equipped two more ships, and the old company joined six to these two. These eight ships set sail together in 1600, again under the command of van Neck.

Unlike the Portuguese, the Dutch took a southerly route across the Indian Ocean, which gave them sound strategic reasons for settling in the Mascarenes. They were pleased with the arable dark volcanic soil of Mauritius. Exploring the island, they noted the rich forests of ebony trees and the abundance of wildlife, including doves, turtles, fish, and flocks of an ugly bird unlike any they had seen before. It was the size of a swan, with a large head furnished with a kind of a hood. It seemed to have no wings, but in place of them it had three or four small black quills and a tail consisting of four or five curled plumes of a gray color. The Dutch called this creature *Walckvogel*, or disgusting bird, because of the toughness of its flesh when cooked for eating, even though they found the pectoral muscles palatable. The ample supply of turtledoves apparently also caused the *Walckvogel* to be the less-esteemed food resource.

> For food the seaman hunt the flesh of feathered fowl
> They tap the Palms, the round-sterne Dodos they destroy,
> The Parrot's life they spare that he may scream and howl,
> And thus his fellows to imprisonment decoy.
> —Jacob van Neck (1598), *Voyage* [15]

According to a French account, Wybrant Warwyck, van Neck's associate, described the dodo in the following terms:

> It is a bird that we call bird of nausea, the size of a Swan, with a round tail, without wings; of these birds we took a certain amount, together with some Turtles and other birds, we boiled this bird but he was so tough that we could not boil it enough, and we ate it only half-baked. We found nobody in the island but we killed a great number of Turtles, since nobody could scare them, they had no fear, stayed in place, and let us beat them. In summary, it is a country with great

Here, van Neck's sailors are shown enjoying the abundant resources on Mauritius. (From Hugh Edwin Strickland, *The Dodo and Its Kindred; or, The History, Affinities, and Osteology of the Dodo, Solitaire, and Other Extinct Birds of the Islands Mauritius, Rodriguez, and Bourbon,* London: Reeve, Benham, and Reeve, 1848. From the collections of the Ernst Mayr Library of the Museum of Comparative Zoology, Harvard University.)

abundance of birds and fishes, more than all others we found during this Voyage.[16]

Admiral van Neck's friend and colleague, Admiral Pieter Willem Verhoeven, took a stroll among the dodos' nests and was "pecked mighty hard" for his curiosity. This episode and others would help form the cornerstone of the lore of the dodo in Europe. Van Neck returned to the Netherlands with four of his ships in 1599. The remaining four ships followed in 1601. The admiral's journal was immediately published in Dutch (1601), French (1601), Latin (1601), English (1601), and German (1602). A flurry of other voyage accounts followed, all sprinkled with tales of a bird that could hardly be real, even as it stood before the amazed eyes of the seafaring men. It was van Neck's expedition, however, that sealed the fate of the dodo, assuring the bird's extinction in less than a century later and at the same time assuring its immortality in the mind of Europe, for van Neck brought live dodos back with him to the Netherlands. It would be the destiny of one of these to become an indelible icon of the modern era.

1 João de Barros, *Terceira Década da Ásia de Ioam de Barros: dos feytos que os Portugueses fizeram no descobrimento & conquista dos mares & terras do Oriente*, Lisbon, Portugal: Ioam Barreira, 1563.
2 *Ibid.*
3 *Ibid.*
4 *Ibid.*
5 *Ibid.*
6 *Ibid.*
7 Hugh Edwin Strickland, *The Dodo and Its Kindred; or, The History, Affinities, and Osteology of the Dodo, Solitaire, and Other Extinct Birds of the Islands Mauritius, Rodriguez, and Bourbon*, London: Reeve, Benham, and Reeve, 1848.

8 Royal decree mentioned in C. R. Boxer's Introduction to
   "Tragic History of the Sea," in: Bernardo Gomes de
   Brito, *The Tragic History of the Sea*, Cambridge, U.K.:
   The Hakluyt Society, 1959.

9 Jan Huygen van Linschoten and Arthur Coke Burnell &
   P. A. Tiele (eds.), *The Voyage of John Huygen van
   Linschoten to the East Indies: From the Old English
   Translation of 1598: The First Book, Containing His
   Description of the East*, London: The Hakluyt Society,
   1885.

10 *Ibid.*

11 *Ibid.*

12 *Ibid.*

13 *Ibid.*

14 *Ibid.*

15 As quoted in: Hugh Edwin Strickland, *The Dodo and Its
   Kindred; or, The History, Affinities, and Osteology of the
   Dodo, Solitaire, and Other Extinct Birds of the Islands
   Mauritius, Rodriguez, and Bourbon*, London: Reeve,
   Benham, and Reeve, 1848.

16 *Ibid.*

Chapter 3

The Emperor and the Painters

How is it that we have a precise idea of what the dodo looked like, down to the shape of the feathers in its wings and the color of the skin around its eyes? There would be any number of written descriptions in the accounts of voyagers passing through the Mascarenes, but these tended to be as sketchy as the woodcuts that would illustrate the printed versions of these narratives.

It was one of the dodos that the Dutch admiral Jacob van Neck brought back with him to Amsterdam that would inspire the most detailed and informative images of the bird. Nothing is known of the details of the transaction—whether it was a gift or a purchase—but one of van Neck's captives was sent from Amsterdam to one of Europe's most powerful rulers—the Hapsburg Rudolf II, monarch of Austria, King of Bohemia and Hungary, and Holy Roman Emperor—at his seat in Prague. Here it would come to the attention of two highly talented painters in Rudolf's court, Joris Hoefnagel and Roelandt Savery, who created lifelike portraits of the dodo. During these times, centuries before the invention of photography, talented painters were best equipped to provide detailed and reliable depictions of flora and fauna, but they gravitated to wealthy patrons and markets; they did not sign on to hazardous voyages to the far corners of the earth in order to witness their subjects first-hand.

One of the most eccentric individuals ever to occupy the Hapsburg throne, Rudolf II (1552–1612) was viewed as feckless when it came to advancing the political-military interests of the lands he ruled as hereditary monarch and the fractious "empire" of German states whose first loyalty was to the Roman Catholic Church. Instead, he is remembered as one of the greatest collectors of all time and for his effort to maintain a center of learning in the tradition of the humanist Renaissance just as the rise of Protestantism and a resurgent Catholicism were splitting Christian Europe into warring camps.

Rudolf II was the eldest son of Emperor Maximilian II and his wife, Maria, the daughter of the mighty Charles V, who for much of the sixteenth century ruled Spain and the Netherlands as well the vast Hapsburg territories centering on Vienna. (He also held the quasi-hereditary post of Holy Roman Emperor.) At his retirement to a monastery at age 58 in 1557, Charles divided his holdings, handing over Spain and the Netherlands to his son Philip II, and giving Austria and other Hapsburg lands, plus the Holy Roman Empire, to his brother Ferdinand I, who was succeeded by his son, Rudolf's father. Born in Vienna, Rudolf spent his boyhood in Spain in the court of his pious uncle, Philip II. Upon the death of his father, Rudolf returned to Vienna to assume his inheritance of the thrones of Austria, Hungary, and Bohemia (the latter roughly corresponding to today's Czech Republic).

In 1576, he was anointed by the Pope to succeed his father as head of the titular Holy Roman Empire of Germany. This "empire" had its origins in conquests in the ninth century by Charlemagne, but over the centuries had devolved into a loose federation of autonomous German principalities some of which, by the late sixteenth century, had become Protestant.

Forsaking the traditional Hapsburg capital of Vienna and its intrigues, Rudolf installed his administrative and residential headquarters in Prague, the capital of Bohemia, where, undisturbed, he could pursue obsessive interests in alchemy,

Rudolph II. A tortured soul and an unlikely leader for the Holy Roman Empire, Rudolf never fought religious enemies but rather generously patronized the arts and the sciences. (Found in Thomas DaCosta Kaufman, *The School of Prague: Painting at the Court of Rudolf II*, Chicago: The University of Chicago Press, 1988. The original painting is believed to be lost.)

astrology, and astronomy. Along the way he became the pro-
tector of the Danish astronomer Tycho Brahe and his German
disciple Johannes Kepler, two men who helped bring about
the scientific revolution in Europe by challenging the notion
that the earth was the center of the universe. The fact that they
were Protestants and Rudolf was supposedly a bulwark of
Catholicism made no difference, since it was not so much
Tycho's and Kepler's enlightened theories as their ability to
do his horoscope that endeared them to Rudolf.

Under Rudolf II, Bohemia became one of the most
advanced countries in Europe in terms of economic development, social diversity, and freedom of thought and religion. It
was not, however, politically and militarily stable. In order to
secure the political and financial support of the so-called
estates—the ruling classes as distinct from the monarchy—
Rudolf allowed them more power to exact concessions in
such matters as operating schools and universities and
naming bishops, in return for their support. This situation
permitted Protestantism to make inroads in populations right
under the Holy Roman Emperor's nose. Not that Rudolf
seemed to care, but his powerful Hapsburg relatives in Vienna
cared, and they set about scheming for his downfall.

As the greatest collector of the age, Rudolf sent his agents
on expeditions all over Europe. The vast museum he created
in Prague went well beyond the Renaissance ideal to become a
landmark symbol of the inquisitive impulse that marked the
era. The emperor was not only interested in antiques, he was
fascinated by contemporary art in all its forms, as well as by
live specimens of exotic plants and animals that the explorers
were bringing to the attention of Europe.

To accommodate his collections, Rudolf transformed the
massive Hradčany Castle overlooking Prague into a gigantic
gallery containing thousands of books, prints, paintings,
sculptures, precious stones, fossils, relics, and curiosities of all
kinds. An inventory of Rudolf's countless drawers and cabi-
nets would show a dizzying array of items including automa-

tons, sculptures made of cherry pits, shells, shark's teeth with gold trimming, crystals, minerals, amber formations, bezoars, two nails from Noah's ark, stuffed ostriches, a stone swallowed by a peasant and retrieved nine months later, an iron chair that closed itself over those who sat on it, a group of deer and gazelles running over an artificial hill, an organ that played madrigals, mandrake roots, cups carved out of rhinoceros horns filled with different poisons, a votive medallion made of clay from Jerusalem, a piece of the clay from the Hebron Valley that God used to model Adam, Moses's staff, lizard skins, beasts of silver, turtle shells, painted wax dolls, Egyptian figurines, mirrors of glass and of steel, trees made of coral, female torsos made of plaster and painted in skin colors, skulls, golden goblets, landscapes made of Bohemian jasper, paintings on alabaster, painted stones, mosaics, glasses of sculpted crystal, topaz bowls shaped like lions, maps, spurs, bits, draperies, decorated weapons, music boxes, even hummingbird wings.

Rudolf also enlarged the gardens of the Hradčany and other imperial residences, and established a botanical garden in Prague, to which his ambassadors and agents sent a stream of rare plants. On the outskirts of Prague, he built a menagerie, or zoo, which he stocked with exotic animals, including camels, elephants, and leopards. The castle boasted fish ponds, a garden for pheasants, a lion's den, filled-in moats where deer roamed (the *fosse aux cerfs*), and an aviary for rare birds, which was presumably where Rudolf lodged the dodo sent to him from Amsterdam. Here, the bird would be on view to some of the most highly skilled artists in Europe.

So it was that but for the manias of this Hapsburg king, entrenched along with a coterie of alchemists, cabalists, necromancers, natural philosophers, astronomers, astrologers, painters, and sculptors in a walled palace stuffed with the most incredible and eclectic collections ever amassed in Europe, we would have little or no idea of what the dodo really looked like. The fact that we do is traceable to the work

At Rudolf's time, the Hradčany Castle housed a fascinating assembly of sages and artists, dwelling both at the darkest and lightest ends of knowledge. Together with the explorers and emissaries the Emperor sent to all corners of the world, these men furnished the interior with the largest and most extravagant collections in all of Europe. For all of these wonderful things, the castle was not an entirely wonderful place, and stories of poisonings, betrayals, and manifestations of sheer madness inside the walls abound in both scholarship and folklore. (Photograph © 2000 Barbara Nabilek, www.experience-prague.com.)

of two artists in Rudolf's employ. One encountered the dodo at the very end of a long, uncontroversial career as the last of the great Flemish manuscript illustrators; the other was an innovator whose sketches of the dodo and other animals and plants would have an important influence on the course of Dutch painting.

The first artist was Joris Hoefnagel, born in 1542 in Antwerp, a prosperous trading city in what is now Belgium, a decade before Rudolf's birth in Vienna. The son of a wealthy merchant family, Hoefnagel traveled in England, France, and Spain as a young man and discovered a talent for painting. Little is known of his early career, which coincided with the opening phases of the prolonged revolt of the Netherlands against Spanish Hapsburg rule. In 1577, after the bloody sacking of Antwerp by the Spanish army, Hoefnagel fled the city of his birth and eventually landed a job as court artist to the Duke of Bavaria. In this post he did his first major work as a miniaturist illuminating a book. He became skilled in depicting major cityscapes that were reproduced as insets in important atlases of the period. In 1591, he moved to Prague to take up an appointment as court artist. There, or in Vienna—the record is not clear—he would die in 1601.

Once under the emperor's wing, Hoefnagel dedicated himself to the naturalistic depiction of animals and plants with exceptional virtuosity. He immortalized the dodo in an oil portrait, a beautiful miniature that at one time was erroneously dated in the period of 1610 to 1626, but then redated to 1600, just when the dodo brought to Europe by Admiral van Neck arrived in Prague.

Three years after Hoefnagel's death, in 1604, Rudolf's other dodo painter arrived in Prague from the Netherlands. His name was Roelandt Savery, and he is credited with five surviving depictions of Rudolf's dodo. Inspired by the sight of the dodo while in Prague, Savery used the bird in paintings made mainly after he left the Hapsburg court in 1613 and returned to the Netherlands. Up until his death in 1639,

GEORGIVS HOVF NAGLIVS ANTVERPIAN
QVI PICTVR: DELICAT GENIO DVCE AMPLEXVS
SPONTE PROMOVIT ET MAXIME ILLVSTRAVIT:
H Hondius fe et ex

Joris Hoefnagel (1542–1601) was one of the first artists to paint still lifes. His miniatures from line subjects, both animals and plants, prepared for a gorgeous compendium called The Four Elements *commissioned by the Emperor, propelled the art of scientific illustration to new and splendid heights. Hoefnagel, a Stoic, also wrote two poems inspired by Albrecht Dürer's paintings* The Hare *and* Stag Beetle, *from which he borrowed two of the images in* The Four Elements. (Courtesy of Staatliche Graphische Sammlung Munich, Germany, and the General Research Division, the New York Public Library, Astor, Lenox, and Tilden Foundations.)

Savery went on painting the dodo over and over, sometimes only from memory. This led to anatomically inaccurate versions, in which the painter went as far as to endow the dodo with duck-like webs between the toes. Exactly which of Savery's depictions were based on sketches he made in Prague of the live bird cannot be determined.

Roelandt Savery was born to a Protestant family in 1576 (the year of Rudolf's coronation as Holy Roman Emperor) in Flanders, in what is now Belgium. He was initially instructed by his brother in the depiction of animals and flowers. In the beginning, he learned to do genre paintings in the style of Gillis van Coninxloo and Pieter Breughel the Younger, village scenes featuring, if not dominated by, domestic animals.

When he was called to Rudolf's court, Savery was given a very specific mission: to travel through the Tirolean Alps and make sketches for paintings that he would execute upon his return to Prague so that the gout-ridden Emperor might appreciate the splendor of his empire without going to the trouble of actually traveling in it. While on Rudolf's payroll, Savery became a master of the Alpine landscape. He also opened a new avenue in the history of art by making monochrome drawings of the denizens of Prague and Bohemia, in which he included his subjects' humble surroundings. These he labeled *naer het leven*, or "from life." The term made observation—with a goal of lifelikeness, as opposed to introspection or ideals—a source for artistic creation. Intended as sketches for paintings, these drawings *naer het leven* often bore written color notations reflecting what the artist actually saw.

Following Savery's example, others of Rudolf's painters moved away from mythology and allegory toward the representation of scenes that were as lifelike and accurate as possible. This "low-life" painting also served as a bridge to the painting of what was indeed regarded as lower, i. e., nonhuman forms of life, such as still lifes, landscapes, and animals. In the process, new conventions as well as new categories of painting came about. At the imperial court of Prague and else-

ROELANT SAVERY

A été un paintre extraordinaire des animaux, et autres oyseaux; et les
paijsages les quelles il faict, sont bien estimées de les amateurs de la painture
il est natif de Flandres, il a esté peintre du l'Empereur Rudolphe second.

Adam Willaerts delin.   Io. Meyssens fecit et excudit.

*Roelandt Savery (1576–1639), a Flemish painter who helped
immortalize the dodo. With time, Savery found a curious
artistic niche, creating huge tableaux depicting Eden-like
scenes of the animal kingdom. These images were increasingly populated by more and more bizarre creatures arriving
at European ports from exotic places. The dodo was included
in some of those paintings.* (Courtesy of Die Kunstsamm-
lungen der Veste Coburg, Germany, Kupferstichkabinett.)

where, up to Savery's time, studies of animals, sea creatures, birds, and insects were done in miniature as manuscript illuminations, culminating in the work of Hoefnagel. But Savery changed all this by making landscape painting, some of it based on studies from nature, fashionable. Savery made an important contribution to the history of Western art by becoming the first important animal painter to work on canvas, creating a genre that would soon burgeon into a major specialty.

Like all painters of his time, Savery needed to find a niche through which he could be recognized, some sort of trademark. For Savery, the niche was his skill at depicting animals. In his paintings, whether in village scenes or landscapes, animals became increasingly prominent, propelled to the forefront as his work progressed. Even though he was destined for fame and favor, Savery apparently was repelled by the malaise permeating Rudolf's establishment up to and after the Emperor's death in 1612. Savery remained in Prague until 1613, then returned to the Netherlands to settle in Amsterdam for three years before moving to Utrecht, where he spent the rest of his life.

From 1610 onward, and with particular intensity after 1618, he became a preeminent portraitist of animals, specializing in paradise scenes and bucolic themes, responding to a new interest by buyers in the details of nature and a new, court-based fascination with exotic animals. Among these was the dodo.

Kurt Erasmus, who wrote a doctoral thesis on Savery in 1908, stated that there were four surviving Savery paintings in which the dodo appears: *The Paradise*, painted in 1626 and now in Berlin; *Landscape with Birds*, painted in 1628 and now in Vienna; *Orpheus*, undated but probably painted around 1626 and now in The Hague; and another *Orpheus*, undated but probably painted around 1628 and now in Pommersfelden, Germany.[1]

In Erasmus's view, Savery's dodo paintings were done when he was a member of the painters' guild at Utrecht, where he moved in 1619, long after his departure from Rudolf's court. Erasmus argues that Savery saw a dodo for the first time only when back in the Netherlands, and never in Prague, since "judging by the paintings, Roelandt Savery has painted this bird from a living example, however not in Prague, but in the Netherlands; for otherwise he would have included this bird already in some of the paintings he had done in Prague, and not only in some of his later works."[2]

This is contradicted by the arguments of Joaneath Ann Spicer-Durham in a 1977 Yale doctoral thesis, and Thomas DaCosta Kaufman in 1988. These two Savery scholars concluded that Rudolf's live dodo provided Savery with his first and foremost inspiration. The matter is complicated by the fact that Savery was not a consistent painter. A number of his drawings have been confused with those of other Dutch contemporaries, including his brother Jacob and even Rembrandt. This is due in large measure to Savery's propensity for exploring a wide variety of subjects, styles, and techniques, both old and new, as well as the absence of signatures and dates.

In the period following Savery's birth in 1576—he was the youngest of four sons of Maerten Savery and Catelina van der Beecke—the Dutch Protestants, led by William of Orange, managed to establish an independent republic in the northern Netherlands, while the Spanish held on to the largely Catholic south. This gave rise to a mass migration within the Netherlands, especially of Protestants. A small number of reformed families fled to England, others to Germany, but most chose simply to move north. In the process, the northern Netherlands and the city of Amsterdam, the center of Dutch sea power and prosperity, attracted the energies of Flemish commercial and artistic talents, both for religious and economic reasons.

The unprecedented growth and enrichment of Amsterdam and the Netherlands in the following half-century, despite an on-going war, was largely due to huge infusions of talent, energy, and money from outside. As producers of nonessential goods and services, artists have traditionally been among the first to suffer in times of economic disruption; at the same time, they can pick up and go elsewhere more easily than those whose livelihoods depend on land and inventory.

Sometime in the early 1580s, Maerten Savery and his family, including two sons, Jacob and Roelandt, went on the move. They first settled in the Flemish town of Courtrai (or Kortrijk), located in present-day western Belgium not far across the border from the French city of Lille.

Though no one knows exactly what the father, Maerten, did for a living, apparently he was not an artist, and therefore did not play a direct role in Roelandt's schooling. This task fell upon the shoulders of Roelandt's elder brother Jacob, a painter considered mediocre by some, who seems to have instructed Roelandt early on in the representation of animals and landscapes. He might have been mediocre, but during the family's flight, Jacob seems to have been the breadwinner, since we find the whole clan locating and relocating wherever he got a job.

After a period of Calvinist domination, Courtrai was taken by Spanish troops in 1581. The Inquisition was not reinstated, but Protestants had two options: abjure or get out. The outbreak of the plague in the same year put additional pressure on the family to emigrate again. After stops in Dordrecht, Antwerp, and Haarlem, Jacob and his family, including his brother Roelandt, reached Amsterdam in 1591, where "Jacques Savery von Cortrijk schilder [painter]" acquired the rights of citizenship. Now fifteen years old, Roelandt would spend the next dozen years studying art in Amsterdam just as that city became the art capital of the world.

Landscape with Jepitha's Daughter *by Jacob Savery. The older brother of Roelandt, and for some years the main bread-winner of the family, Jacob initiated his sibling in the art of drawing landscapes.* (Courtesy of Rijksmuseum Amsterdam.)

During the closing decade of the sixteenth century, Amsterdam harbored a flourishing Flemish community, including a great number of artists whose pivotal role in the rapid achievements of the Dutch school of painting has been compared to that played in New York by French émigré surrealists after World War II. The overwhelming majority of these Flemish artists, among them Jacob and Roelandt Savery, were Protestants who painted landscapes (with a few genre painters among them). Some, if not most, consciously rejected devotional themes in keeping with the Protestant antipathy to explicitly religious "images," preferring instead the lessons of earthly life and the revelations of nature. At the time Jacob acquired his citizenship, Amsterdam painters were becoming known either as romantic landscapists or naturalists, or both.

It should be noted that tolerant Amsterdam didn't have huge collections of art, but instead possessed an abundance of small private collections, assembled by amateurs and artists themselves, of drawings and prints, as well as paintings by Flemish masters brought north by immigration and trade.

Roelandt Savery's artistic activities during this period leave much to supposition. It is known that in 1600, he began producing dated paintings of animals, flowers, and landscapes, and a few drawings. His brother Jacob remained his most important influence, although van Coninxloo and Breughel the Younger were conscious models, as well. Then Jacob died from the plague in 1603. The possibility that the young painter made visits to Paris and Rome is unsubstantiated, but it is definitely known that Roelandt Savery received the invitation to go to Prague in 1604. This must have reflected suppositions on both sides. Savery must have achieved something of a reputation as a painter of animals and landscapes, for as soon as he arrived in Prague, Rudolf sent him to the Tirol to produce grandiose alpine portraits. For his part, Savery must have been attracted enough by the offer to leave Amsterdam and its excitement. Possibly the reputation

and the size of Rudolf's collections were factors in his decision.

It is hard to tell what artistic influences, if any, Savery was
exposed to in Prague. While Rudolf's patronage had made
Prague an important artistic center by the time of Savery's
arrival, there never was a "School of Prague" in any conven-
tional sense. Local painters and sculptors did not receive the
favor of the court, and the painters' guild of Prague had no
power over the foreign artists, who were employees of the
court and apparently enjoyed a form of diplomatic and artistic
immunity. The major artists who resided at the court were, for
the most part, already formed upon arrival. They constituted
an assemblage of individual talents brought together by the
bidding of one man, and their efforts were largely absorbed by
that one man. Even though as Holy Roman Emperor, Rudolf
was the duly anointed champion of the Catholic Church in
Europe north of the Alps, he made little effort to foster the
goals of the Counter Reformation through the visual arts, preferring the conceptions of "history painting" for his own
pleasure and glorification. The "history" painters recruited
by Rudolf worked in the mannerist style and, wholly
beholden to the whims of their patron, did their work in a vir-
tual vacuum. Savery's awesome views of the mountains of the
Tirol and the great valleys of Bohemia, however, were based
on studies "from life," made at Rudolf's behest to evoke the
magnificence and variety of his domain. As such, they procured Rudolf's satisfaction in more metaphoric terms than
their heightened realism suggests.

Savery's ten years in Prague—including the two years he
spent on his sketching trips in the Alps—were the peak of the
artist's activity and achievement as a draftsman and painter.
Over 50 of his paintings can be traced to this period, as can
the majority of his datable drawings, including all his peasant
drawings. Savery also produced a large number of drawings of
Prague itself (even Rembrandt didn't make so many of his
beloved Amsterdam), which are valuable contributions to an

understanding of the cityscape under Rudolf. He continued to paint "bouquets" (paintings of flowers) for all these years, and made studies in Rudolf's menageries for his glorious pieces of Edenic animal kingdoms realized after his return to the Netherlands.

In one respect, Rudolf's vast collections functioned simply as instruments of imperial policy, to awe foreign diplomats with their incomparable magnificence. On another, more spiritual level, the collections represented a microcosm of the world in its infinite richness. In this respect, Rudolf's fascination with science, pseudoscience, and metaphysics can only be fully appreciated as fitting into his age's systems of knowledge.

As Spicer-Durham points out, late-sixteenth- and early-seventeenth-century encyclopedic collections constituted an intellectual pursuit, an attempt through the visible, viable microcosm embodied in an assemblage of artifacts to reflect and comprehend the invisible macrocosm in all its variety:

> The functioning of a coherent cosmic hierarchy linking all objects and processes implicit in the belief in a perpetual world soul was sought not simply in the data of empirical observation but in the divination of underlying unitary, universalistic systems and necessary truths which once recovered might be used to organize error-ridden human perceptions and knowledge. Linking all matter in a "chain of being" was construed as the surest sign of God's hand as manifested in His creation, the visible world. It is the apprehension of God the Creator that justifies the empirical sciences, and the urge to apprehend which motivates the "scientist." The ordering of rarities as well as more representational objects within a collection in terms of the realms of nature and art, God's creation unaltered or in a state altered by man—naturalia and artificialia—is therefore a prime manifestation of this contemporary preoccupation with the general ordering of the phases of human knowledge, with the attempt to confirm a priori relationships both scientific and metaphysical.[3]

The Renaissance preoccupation with unitary circles reached its apogee in the magical universe created by Rudolf. His collections and the studies he commissioned of natural phenomena expressed the same impetus that animated others who enjoyed his patronage at the court of Prague: the alchemists' pursuit of the philosopher's stone, the cabalists' striving after the divine, or the theses of the occult Hermetic philosophers. All participated in a grandiose effort to decipher Nature's secret codes. In this light, Kepler's search for the harmony governing the motion of the planets may be seen as a repercussion of the assertions of harmony in Giuseppe Arcimboldo's celebrated "portraits" made up of fruits, flowers, and animals. Similarly, Hoefnagel's representation of natural creatures corresponded to a view of the natural world as full of divine order or hidden meaning, like a system of hieroglyphs.

As part of their quest, many court painters keenly sought the "miraculous" when depicting natural creatures. They placed heavy emphasis on such things as deformed antlers, abnormal growths, gigantic beetles, strange creatures found in rocks—all manifestations of Nature's capacity for defying norms and breaking molds. This expressed a line of reasoning by which it was in the exception, not in the rule, that the deeper meaning of things can be found. So it was that Savery's first mission for Rudolf was to seek out and draw rare wonders of Nature for later use in paintings. Savery's alpine drawings were not predictable pleasant vistas, but rather rugged landscapes filled with rocky crags and valleys, waterfalls, strange rock formations, and precariously leaning trees—wondrous things indeed.

Working "from life" thus became a means of probing into life's subterranean text by noticing in life subjects what immobilized, idealized figures in books could never give away. Hoefnagel sometimes pushed this preoccupation to the point of attaching the wings of his insect models to the pages next to his paintings of them. It was the same impetus that motivated

Rudolf's artists to work outdoors to sketch even single land-scape elements such as trees. Savery's journey to the Alps is one of the first examples in art history of a commission to create independent landscapes specifically taken from nature, to make drawings outdoors to be used in paintings.

Savery's trip to the Tirol and his later excursions around Bohemia demonstrate that a new impulse had entered the painting of landscape. While Savery never painted an exactly identifiable setting, the motifs that he used can often be local-ized. For instance, Savery would sketch a completely crooked tree without any leaves in his notebook, but when the tree appeared in paintings, it was often put in yet another, even more awesome Alpine landscape. Thus his work can clearly be distinguished from that of his predecessors, in which both the exact location and the specific motifs are totally imaginary.

In addition, during Savery's time, the depiction of animals reflected the simple and venerable notion of man's God-given responsibilities and privileges of dominion over the earth, as established in a number of biblical passages. In Genesis, after the Flood, God tells Noah:

> The fear and dread of you shall fall upon all wild animals on earth, on all birds of heaven, on everything that moves upon the ground and all fish in the sea; they are given into your hands. Every creature that lives and moves shall be food for you; I give them to you, as once I gave you the green plants.[4]

Similarly, in Psalm 8, it is said that:

> Oh Lord our sovereign,
> How glorious is thy name in all the earth!
> Thy majesty is praised high as the heavens,
> Out of the mouth of babes, of infants at the breast,
> Thou hast rebuked the mighty,
> Silencing enmity and vengence to teach thy foes a lesson.
> When I look up at thy heavens, the work of thy fingers,
> The moon and the stars put in thy place by thee,
> What is man that thou shouldst remember him,
> Mortal man that thou shouldst care for him?

Yet thou has made him little less than a god,
Crowning him with glory and honour.
Thou markest him master over all thy creatures;
Thou hast put everything under his feet:
All sheep and oxen, all the wild beasts,
The birds in the air and the fish in the sea,
And all that moves along the paths of ocean.
Lord our sovereign
How glorious is thy name in all the earth![5]

Both passages, in turn, involve the notion that close contemplation of the creatures of the earth helped apprehend their Creator and learn to discern the virtues and the wrongs that the Creator distributed universally among all His creations. Growing out of the artistic tradition of graphic illustration, painting on canvas was enhanced in its appeal by the association that the educated viewer might bring to the experience, so that the paintings could be perceived and enjoyed on more than the aesthetic level alone.

Literary fashion in the period was similar. In the sixteenth century, writing about natural history and hunting, and commentaries on Aesop's fables in verse or prose, were hugely popular. Like pictures of animals, stories about them helped expand contemplation of the natural world and thereby glorified and satisfied the Creator. The discovery of previously unknown creatures by European voyagers in far-off corners of the world, instead of challenging the Christian model, simply magnified God's greatness, representing a recovery of things lost to men since the expulsion from Eden.

As the sixteenth century advanced, an increasing number of illustrated natural history texts was published, ranging from new editions and rehashings of the ever-popular works of Albertus Magnus and Bartholomaeus Anglicus to new classics by Konrad Gerner and Pierre Belon. The growth of artistic interest in God's creatures reflected a spirit of curiosity that permeated the observational sciences of the period. In painting, this was expressed in the choice of a zoological collection or garden as subject, its occupants selected and arranged to stimulate the viewer both to remembrance and discovery of God as well as innocent delight—as far as possible serving as a kind of memory theater for the exercise of one's mental capacities.

Insofar as painters of the fifteenth and sixteenth centuries sought to imitate the natural world—and in so doing rival the creative powers of Nature itself—they shared with contemporary students of plant and animal life a curiosity about Nature and the desire to go beyond mere commonsense experience. Further bridging art and science in Savery's day, some of these pieces, especially those of the so-called "birdyard genre," might have resulted from the request of, say, a naturalist in Amsterdam or follower of the pioneering botanist Charles de L'Ecluse (better known by his Latin name, Carolus Clusius) for an assemblage of natural history studies, both exotic and native. These renditions would be more artfully conceived than the relatively sterile engraved series or picture books or isolated studies in gouache then in circulation. An ideal bird garden brought almost to life in the "birdyards" was an idealized garden or park that the naturalist, the poet, the collector, or any other thoughtful soul could enter and read from the scene whatever he wished.

Another form of animal painting in which Savery excelled was the so-called "peaceable kingdom," based on the prediction of Isaiah of what will happen when the Messiah comes. These paintings were depictions of predator animals and their prey—the wolf and the lamb, the lion and the gazelle, resting side by side—animals that would never coexist peacefully in the wild. More often than not, a human being like Orpheus or Adam, or God Himself, was there to preside over this unearthly harmony.

The only peaceable kingdoms attributable to Savery in the Prague period are presided over by Orpheus (a series of at

One of Savery's "birdyard" paintings. This series grew increasingly complicated in its mixture of animals, colors, postures, and backgrounds, challenging the painter to compile more and more information on exotic fauna. Some of the creatures were drawn from life, and others from a profusion of natural history manuals circulating at the time. (Courtesy of the Koninklijk Museum voor Schone Kunsten, Antwerp, Belgium.)

least five paintings), but this could very well reflect his patron's tastes and visions. It is not difficult to imagine that the figure of Orpheus, archetypal poet and the Renaissance embodiment of human culture and harmonious rule, through whose magic and wisdom natural hostilities were suspended in an explicit example of cosmic harmony, would be extremely attractive to Rudolf, who fantasized himself as a peacemaker between Catholics and Protestants. In fact, depictions of Orpheus by other painters abounded in Rudolf's collections.

And yet, the magic spell of Orpheus appears as an afterthought in Savery's paintings. The first composition of the series depicts the animals presenting themselves two by two, an arrangement suitable for waiting before the Ark but totally inappropriate in the story of the bard. Also, Savery has his Orpheus charming both wild and domesticated animals, which runs counter to the basic version of the myth in which wild animals responded to Orpheus playing his lyre. These anomalies suggest that Savery had worked on his peaceable animal kingdoms with other themes in mind prior to introducing Orpheus. Indeed, in most of these animal gatherings, the animals seem to pay Orpheus very little attention. As if to commemorate his release from Rudolf's obsessions after departing from Prague, in a later painting Savery broke the prevailing harmony of the peaceable kingdom by showing the stoning of Orpheus by the Thracian women, while a first sign of turbulence ripples through the assembled beasts.

Savery's earliest renditions of animal themes, done between 1600 and 1601, prior to his going to Prague, contained birds and beasts in poses destined to recur throughout his work. For instance, a monkey, which first shows up on the back of a dromedary when the animals are being introduced to the Ark, jumps from scene to scene through subsequent paintings without ever changing its posture. Savery must have intended from early in his career to build a repertoire of drawings of models taken "from life" in order to support the

This Savery painting shows Orpheus enchanting wild animals with his lyre. Mixing in one single scene several animals that would never cohabit peacefully in real life became one of the painter's most celebrated trademarks. It is quite probable that this approach was developed in response to a request from Rudolf, who dedicated substantial attention to the behavior of different wild animals after they were locked together inside a fence. After leaving Prague, Savery never returned to this theme. (The National Gallery, London.)

remarkable "lifelike" quality of his creations. But this does not mean he actually depicted the animals *from life*.

How, for instance, could Savery have studied the water buffalo in his *Garden of Eden* while in Amsterdam? Birds and beasts were maintained at the Amsterdam *Dolhof*, or pleasure garden, but were limited in number and variety; and there was no princely menagerie accessible to Savery in Amsterdam. The royal court of the House of Orange at The Hague had no zoological collection of significance at this time, and Clusius's renowned botanical garden at Leiden had no zoological counterpart.

On the other hand, even before the chartering of the East and West India Companies (1602 and 1609), sea captains and traders exhibited exotic birds and animals as valuable curiosities in the Netherlands' chief port. Among the first of these was a live cassowary, a tall, ostrich-like bird native to Australia, New Guinea, and other parts of the East Indies, which was by no means an endangered species and would often be mistaken for the dodo in nineteenth-century analyses of seventeenth-century prints. It arrived in Amsterdam in 1597 with the return of the expedition under Captain Cornelis Houtman. In 1605, Clusius reported in his *Exoticorum libri decem*, published in Leiden, that this bird was sold to Count van Solms, a prominent Dutch nobleman in The Hague, who eventually gave it to the Archbishop of Cologne, who, in turn, presented it to the always-eager Rudolf. But we have no idea whether the bird was dead or alive during all these transactions. Conceivably, Savery might have seen the cassowary twice, once before his departure from Amsterdam and a second time in the collection of Rudolf. But judging from his rendition of the animal, it is easier to reason that he got the shape and color from prints by others.

Once he was in Prague, Savery was able to draw on numerous sources for the animals in his repertoire: ancient and contemporary art, both paintings and sculpture, collected by Rudolf, as well as live and preserved specimens. There are indications that he had access to early-seventeenth-century Islamic miniatures of birds presented to Rudolf by Persian ambassadors.

Savery's predecessor, Joris Hoefnagel, had illuminated four books for Rudolf—one devoted to four-footed animals, a second to creeping animals, a third to birds, and the fourth to fish—and received a thousand gold coins for them. These four exquisitely emblematic books were named after the Four Elements, and, in the spirit of Rudolf's establishment, constituted an ideal catalog of God's animated handiwork comprehended under the cosmic authority of the four elements. Stringent realism in the representation of the myriad facets of Nature animated the universal, elemental allegory of the whole. So compelling was the realism of these books that Hoefnagel came to be considered a naturalist in his own right. In a 1607–11 catalog, he was cited as an authority on the Latin names of fish, alongside the celebrated Clusius. And it is generally accepted that Savery used Hoefnagel's compendia exhaustively for his flower drawings.

Nearly all the animals in Savery's paintings up to at least the mid-1620s were apparently derived from studies made of either living or embalmed specimens, and many of these animals pop up time and again in paintings and drawings as the years pass, never losing their precision. By various means, Savery had built himself an impressive paper menagerie by 1601, which was enriched during the Prague period by the wealth of materials in Rudolf's collections. It is interesting to note that deer appear more consistently in Savery's paintings than any other animal, except for cows in later years. The stately stag, prerogative of the royal hunt, is represented by far more stuffed or sculpted heads than any other animal in the inventory of Rudolf's collections made in Prague in 1621.

Besides the many studies he made of lions and horses in Rudolf's service, Savery sketched leopards, bears, foxes, and a wolf, and many exotic birds. Rudolf had a pet eagle presumably kept in regal surroundings, and journalistic reports from this period mention several eagles behind fences. They do not mention, however, the sight of cassowaries, flamingos, redtails, or dodos—all newcomers to Europe and all drawn convincingly in a later work titled *Museum des Kaisers Rudolf II.*

It is clear that Savery developed a strong feeling for the dodo's waddling gait. This is demonstrated by his Prague sketches of the bird in several postures, which suggest he was well acquainted with the dodo in life. Since the only dodo documented in Prague was stuffed at least by 1611, it is likely that the different dodos, including the live bird studied by Hoefnagel for his own dodo painting, were in fact the same luckless creature.

On the other hand, the models for the armadillo and the crocodile Savery used in sketches for such works as his 1620 *Garden of Eden* most likely were stuffed specimens, among those listed in the 1621 inventory of Rudolf's Prague possessions. Likewise, the parted, snarling jaws of the *Boar at Bay* may have been calmly scrutinized indoors, since such trophy heads are listed in several inventories. Alternatively, the only existing Savery sketch of a boar's head was based on a severed boar's head on a trencher.

Another big attraction of Rudolf's domain for an animal painter would have been stuffed birds. For one thing, their immobility made them ideal for studying. Indeed, some of the birds winging their way through Savery's late paintings really look as if a perch or mount has been airbrushed out from under them, while others appear glued to the ground, at times frozen in dramatic but artificial poses.

Yet again, it is most likely that Savery had no personal acquaintance with the rhinoceros, live or stuffed. Therefore, the phantom appearance of a rhinoceros occasionally plodding through the background in several assemblies, starting with the first rendition of *The Animals Before the Ark*, must have been derived from prints. The heavily armored but watchful rhino in his 1617 *Orpheus*, for instance, seems to be

This detail of the exotic wildlife tableau The Fall of Adam *is one of Savery's most famous dodo portraits. Although dodologists long debated the true authorship of the painting, art critics unanimously attribute it to Savery.* (From Hugh Edwin Strickland, *The Dodo and Its Kindred; or, The History, Affinities, and Osteology of the Dodo, Solitaire, and Other Extinct Birds of the Islands Mauritius, Rodriguez, and Bourbon*, London: Reeve, Benham, and Reeve, 1848. From the collections of the Ernst Mayr Library of the Museum of Comparative Zoology, Harvard University.)

A bird lived long ago
Near the coast of Madagascar.
She'd call herself Dodo
If you ever thought to ask her.
For the birds all had ways
In the good old days
Much like the ways of men:
They could speak, they could laugh,
They could flirt, not half,
From their roost to their roost again.

Historians agree
That the dodoes were a bevy
As good as good can be,
Though perhaps a little heavy.
And their fat made them calm,
They would do no harm,
And foes they had none to dread,
So they waddled about,
Ever in and out—
What a wonderful life they led![6]

modeled on a large illustration by the sixteenth-century artist Konrad von Gesner.

In the inlet to the rear of an *Orpheus* dating from Savery's early Prague years, there is a glowing sperm whale, ferociously spouting. Taken together, its greatly enlarged eye and a flipper transformed into an ear convey a striking countenance. This rendition could have come from Savery's own memories, since before going to Prague he may well have seen beached whales in the Netherlands, either at Beverwijk in 1598 or at Noordwijk aan Zee in 1601. Alternatively, Savery's model could have been Jacob Matham's widely circulated (and copied) engraving after a drawing by the painter Hendrick Goltzius of the whale beached at Beverwijk in 1598. In his engraving Matham repeated the mistake of Goltzius in turning the whale's flipper into an ear.

The plants depicted by Savery in his idyllic scenes are as arbitrarily selected as the animals themselves. In one Eden, an elephant rubs against a deciduous European tree, while in another the setting consists largely of palm trees. Consistency was never Savery's strong point, and composition was more important than realistic environments. In the Eden with palm trees, Savery's preoccupation with creating the image of a tropical habitat goes as far as decorating the shores with all kinds of exotic shells (items in great supply in Rudolf's collections), and he seems to have adapted scenes from contemporary travel books, such as Theodor de Bry's *India orientalis*, published in Amsterdam in 1601. That Savery's palm trees appear to bear pinecones could be explained as a misinterpretation of the nut clusters in de Bry's work.

Numerous repetitions of the same animals in exactly the same poses from scene to scene add another characteristic to Savery's depictions: humor, even caricature. In one painting, we first see only one cow, and then notice that there is second cow's head resting in bovine sympathy on the back of the first, with no body attached. This kind of incompleteness is frequent in Savery's mature art and unique to him.

One of Savery's studies for the *Garden of Eden* is a light-hearted portrayal of an Indian elephant completely absorbed in relieving an itch of difficult accessibility. Treated with casual familiarity, the noble beast is caught in an undignified posture that invites sympathy and emphasizes the animal's girth. This easy familiarity with less than heroic moments in an animal's daily life is found throughout Savery's work, and would be imitated by animal painters in generations to follow. Note that this elephantine itch is reserved for drawings alone—scratching does not appear in the paintings. Even though in paradise one sometimes has to scratch, decorum is never violated in Savery's kingdoms.

In Savery's studies of camels, the animals' features evolve more and more toward caricature as the chalk sketches progress. The final drawing in a series, *Two Camels*, demonstrates an intimate familiarity with the species, which allows the artist to go beyond sober realism to the light humor of distortion. This curious touch is typical of the series, probably from the end of Savery's stay in Prague.

The same applies to Savery's heavy-rumped dodos from the drawing titled *Dodos*, finished some time between 1610 and 1618. These two dodos, captured in what Spicer-Durham describes as "characteristic but ignoble postures," are trademark examples of specimens truly drawn from life, as opposed to prints, stuffed specimens, or sculptures. Their proportions are exaggerated—intended caricature—and executed with an ease that makes us certain that they were, at some point, moving before the artist's eyes. The outsized head and the large rump that later became standard in dodo cartoons are enlarged in proportion to the rest of the body, and the smile evokes what Spicer-Durham calls "the Dodo's famed guileless stupidity," a forerunner of the persistent image of the stupid, fat bird doomed to extinction.[7] These *avant-la-lettre* dodo cartoons occupy a distinct niche in the painter's career, along with his caricatures of elephants and camels. It is likely that, from sketches done while working on this dodo

Roelandt Savery's Elephants and Monkey. *In the drawings and sketches included in this series, as in many other cases of animal depiction, the painter starts verging toward caricature, exaggerating the most prominent features of the subject, or representing it in funny poses. For instance, one of the elephants in the sketches is scratching itself against a fence.* (Crocker Art Museum, E. B. Crocker Collection.)

caricature, Savery evolved his model for the mature, fat, totally static specimen that appears in the 1627 *Orpheus*.

———————

Unfortunately, for all his scientific and artistic insight, Emperor Rudolf II was a deeply troubled man who inherited a strong share of the Hapsburg family's tendency to insanity. Early in his rule he became prone to severe bouts of depression. He never married, much as his mother begged him and his political allies and consultants pressed him. Not only was he fearful of an heir who would try to steal the throne from him, he was also painfully aware of the evils that consanguinity brought upon the Hapsburgs. Stories of dementia among his ancestors abounded, and during his lifetime he had striking examples close at hand to remind him of the family's curse. As legend has it, one of these was none other than Elisabeth, Countess of Nadasdy, who may have been Rudolf's sister and was reportedly a vampire who bathed in fresh human blood to keep her youth. She was walled into her castle by angry villagers after the discovery of some 600 drained female cadavers in the cellar. Rudolf was so shocked by this tale of horror that he was all the more determined not to perpetuate the chain of family madness.

Initially he was content to keep the company of "good women," which did not prevent him from contracting syphilis. Then he fell in love a with a beauty named Catarina da Strada, the daughter of one of his most devoted and successful art dealers and agents, the antiquarian Jacopo da Strada, after seeing her in a Titian portrait he acquired. Rudolf promoted her father to the post of superintendent of the imperial collections and made her his mistress, and by her he had six children. He was pleased enough by the birth of his first son, whom he christened Julius Caesar, that he briefly entertained the idea of legitimizing him, but stopped short after studying the horoscopes drawn on the day of the child's birth, which were catastrophic.

Epileptic from birth, Don Julius grew up to be mean-spirited and perverse, eventually forcing Rudolf to exile him to a distant castle. Here Julius devoted himself to hunting, skinning, and stuffing the animals he killed. After he raped and hacked to pieces the beautiful daughter of a barber, servants found Julius naked, covered with blood and excrement, howling and kissing the pieces of his victim. The servants sealed him in a room and reported to Rudolf that his son had killed his mistress during an epileptic fit. Soon afterward, Rudolf was told that his son had jumped through a window to his death.

From the very beginning of his move to Prague, Rudolf failed in his efforts to confront the Turks, and this failure plus his increasingly eccentric behavior led to a near-paralysis of Hapsburg power. Once installed in Prague, Rudolf left Austria to be ruled by his brother Ernest as governor. Ernest died in 1595 and was succeeded by another brother, Matthias. Starting in the year 1606, when he was formally recognized as head of the family, Matthias, backed by other Hapsburg relatives, set about bribing the "estates" of Austria, Hungary, and Bohemia to recognize him as king instead of Rudolf. For several years, the Bohemians remained loyal to Rudolf, taking advantage of the situation to extract from him the so-called Letter of Majesty in July 1609. This granted Protestants full freedom of worship and assembly, and delivered control of some aspects of the schools and universities of Prague to the estates. While this development put Prague at the forefront of Europe in terms of religious tolerance, it smacked of republicanism to the Hapsburgs of Vienna. When Rudolf, in 1611, bungled an attempt to suppress the Bohemian estates, they switched their allegiance to Matthias.

By this time Rudolf was often drunk or semi-comatose. He had grown fat and weak; his fingers and legs were badly deformed by gout. He had a false chin to disguise the ravages of syphilis and a toupee to cover his baldness. Unopposed, Matthias invaded Bohemia and marched on Prague. Ambassadors, ministers, and priests fled the palace leaving Rudolf with a faithful group of artists and astrologers, Johannes Kepler among them. Not only did they love the Emperor who had always supported them with ravishing grace, but they had nowhere else to go. Bed-ridden and in a coma, when Rudolf came to, he refused to leave his chambers. Matthias had himself crowned and expelled Rudolf from Hradčany Castle, confining him to the Belvedere Summer Palace.

Years before, Tycho Brahe had told Rudolf that his pet lion, the same lion to whom the zookeepers once threw a drunken barber-philosopher who had fallen out of the Emperor's grace, would precede him in death by only a few days. Now, early in January 1612, the beast died at his master's feet. That night the legs of the dethroned emperor became so swollen that no one was able to remove his boots. Two days later, doctors cut away the emperor's boots, exposing legs riddled by gangrene. His screams were said to echo as far as the palace's front doors. In his final moments, he stopped screaming and ordered the priests and doctors out of his room. He died on January 20, 1612, at age sixty, with only two chamber-boys by his side.

After Rudolf's death, Roelandt Savery remained in Prague, now in the service of Matthias. According to Hapsburg archives, in February 1613, Matthias requested the Bohemian estates to pay moneys to "our *camermaler* and loyal, dear Roelandt Savery" for a three-month trip to Amsterdam.[8] In the following June, the new Emperor directed the estates to pay Savery 300 guilders in partial payment of his outstanding stipend. This implies that Savery's departure to Amsterdam depended on the receipt of the funds and that he was expected to return to Prague. The reason for the trip seems to have been the need to settle his dead brother Jacob's estate and make related financial dealings on behalf of Jacob's children. When exactly Savery left is unknown; but the Hapsburg archives indicate he never returned despite repeated efforts by Matthias.

In Amsterdam, the predominant painting style was that of the Romanists, which Savery ignored. Instead, he continued to elaborate on his proven formulas for landscape and animal paintings. He now turned his attention to the barnyard— embarking on what would later become a major theme in Dutch painting. He also began producing his post-Prague animal and landscape drawings, including the views of Amsterdam.

In 1619, Savery moved to Utrecht, where he spent the rest of his life. According to his contemporaries, he worked only in the mornings, spending most of the resulting leisure time in "merry company," frequently accompanied by his nephew Hans Savery. Over the next ten years, Savery prospered, capitalizing largely on his landscapes and animal paintings— which were indeed highly prized, as is evident by the 700 guilders paid in 1626 by the City of Utrecht for one of his *Paradise of Animals* tableaux. This was presented to Amelia van Solms at her marriage to Frederick Hendrik, the successor to Maurice of Nassau as Prince of Orange and ruler of the Netherlands. Incidentally, in 1626, another dodo was brought to Amsterdam from Mauritius, which may explain why the dodo figures so prominently in the 1626 painting by Savery known as *The Paradise* (now displayed in Berlin). Savery also sold paintings in 1628 to the Prince of Liechtenstein, and in 1637 to King Charles I of England.

The popularity of Savery's work in the Netherlands was partially a reflection of the country's newly acquired colonial power and wealth. Savery's success was also linked to the publication of all sorts of voyage tales and the ongoing revival of interest in Aesop's fables, of which a number of translations and variations were published in the Netherlands between

1617 and 1633. Although Savery does not seem to have depicted the fables himself, the level of interest in them may have stimulated demand for Savery's own animal paintings.

Between 1617 and 1620, Savery produced a cluster of fine paintings of the animal kingdom, more densely populated than ever and increasingly removed from the austerity of the naturalist's study. However, his thematic preferences were no longer concerned with daily life beyond the pasture and the farmyard, except for an occasional trip to the market. Also, while Savery was in Utrecht, his career as a "florist" reached its culmination in an exquisitely fresh but subtly orchestrated "Bouquet" dated 1624. The mountainous terrain of the alpine drawings now softened into something less awesome and more accessible, even as the original, rugged vistas were copied or adapted for decades by a variety of Dutch artists. Although he must have gradually lost interest in drawings as a vehicle of expression, he incorporated studies of picturesque Dutch or "Roman" ruins into his animal paintings, creating an odd sort of pastoral.

Some of these later paintings are of high quality, but the bulk, although signed R. SAVERY, were executed with only minimal participation by the master himself. Some of these less carefully crafted pastiches were the work of Hans Savery, Roelandt's nephew and the only student he was definitely known to have taken on. Still, he had collaborators, namely figure painters, and he probably at times engaged some assistants. It is clear that between 1620 and 1630, a good number of Utrecht's most gifted younger artists were influenced by Savery's drawings and paintings, both his late work and that from his early years.

It is tempting to imagine the man who left us the colors and the looks of the dodo in lively company, dying of old age, rich and fulfilled. But in a eulogy written in 1639, one year after the painter's death, by a somewhat mysterious figure named W. Rogman implies that at the end of his life, Savery was no longer in full command of his reason. Other accounts are more blunt: Savery was poor and demented when he died. *Sic transit gloria mundi.*

———————————

Something similar may be said for Rudolf's collections, although their scattering took centuries and still may not be complete. The looting probably began while Rudolf's corpse was still warm. The story goes that he had discovered the elixir of eternal life and youth, which he kept in a bottle on a ribbon around his neck. While his body awaited burial, a priest snatched the bottle and drank the elixir. He was arrested, and hung himself in prison.

This did not deter others. A certain Jeronym Makovski had been in charge of Rudolf's wardrobe and had on occasion borrowed his master's finest clothes. He would then pass himself off as Rudolf, charging a heavy fee for granting audiences to gullible subjects in the semi-obscurity of the Palace's stables. Makovski was able to buy his freedom from jail by offering to reveal to his jailers the location of many of Rudolf's endless hidden treasures. This was nothing compared to what would follow.

The serious damage to the collections started after Rudolf's successor Matthias, also childless and as weak a leader as his brother, agreed to turn over the crown of Bohemia to his cousin, Ferdinand II, in 1617. On May 23, 1618, Protestant members of the Bohemian estates, furious at the refusal of Ferdinand to respect the freedoms granted to them by Rudolf, threw two imperial governors out of a window of the Hradčany, an event ever since known as the Defenestration of Prague. The victims landed on piles of garbage and were unharmed, but their fall was the pretext for all kinds of art objects, used as mere projectiles, to fly after them through the windows and crash to pieces 50 feet below.

The first inventory of Rudolf's collections was made in the following year, 1619. The looting having then barely started, over 3000 paintings were listed, including works by Michelangelo, da Vinci, Raphael, Giorgione, Dürer, Holbein, Cranach, Breughel, Titian, Tintoretto, Veronese, and Rubens—together with 2500 sculptures and thousands of objects valued at the previously inconceivable sum of 17 million guilders.

The Defenestration of Prague triggered the Thirty Years' War, which in its first phase was a struggle for Bohemia, where Ferdinand employed an army under the command of his cousin, Duke Maximilian of Bavaria. Defeating the Bohemian rebels at the battle of White Mountain on November 17, 1620, Maximilian decided to pay himself for his help. He loaded 1500 wagons with works of art and precious objects from Rudolf's collections. It was said that such a convoy of riches had not been seen on earth since the Queen of Sheba brought her treasures to Solomon.

Eleven years later, in 1631, as the Thirty Years' War dragged on, the Elector Prince of Saxe occupied Prague and took home yet another 50 wagons filled to the top with treasure. On July 26, 1648, Count von Königsmark took over the fabulous Hradčany Castle. From the booty destined for Queen Christina of Sweden, he picked out five wagons overflowing with gold and silver for himself. After the Thirty Years' War, a commission charged with assessing the state of the castle found only broken statues and empty frames. But this was a hasty estimate. In 1749, Empress Marie-Therese redecorated the castle, and thus erased for good the vestiges of the settings where Rudolf once lived. At this time, she sold several paintings to the Dresden picture gallery.

But she couldn't have sold everything because when Frederic II of Prussia directed his cannons against the castle, he destroyed an amazing number of precious objects. The frightened servants trapped inside stored all they could inside huge boxes and hid them in caves to protect their contents. As walls were shaking and chandeliers were falling to pieces, the

desperate flight of those still inside the castle led to the breaking and scattering of hundreds of delicate pieces in porcelain,
crystal, and marble.

On May 14, 1788, when Joseph II decided to transform the
old castle into military barracks, he organized an auction of its
superfluous contents. Before the auction took place, Joseph
ordered that all objects considered of no value be dumped
into the moat. Statuettes, coins, shells, fossils, medals, stones,
and plasters thus sank ingloriously into the mud. Forty years
later, they were still objects of treasure hunts by the street kids
of Prague.

In Joseph II's auction, the famous *Das Rosenkranzfest* by
Dürer, a painting that Rudolf had ordered be carried on men's
shoulders across the Alps as a holy relic, was given away for a
few coins. The statue of Ilioneus, for which Rudolf had paid
ten thousand ducats, was sold for a handful of coins.

Still the treasures were not entirely depleted. In 1876, an
inspector sent by Vienna found more paintings, and organized discreet transfers between one capital and the other. Even
as late as the Nazi occupation of Czechoslovakia (1938–1945),
there were still riches to loot, as evidenced by a steady stream
of crates to Germany. SS Reichsprotektor and Obergruppen-
führer Reinhard Heydrich ordered the crown of Wenceslas,
patron saint of Bohemia, walled up. In his post-war trial, Karl
Hermann Frank revealed to the court the location of this monumental treasure, which otherwise would still be lost in the
underworld of the castle. And God knows what else is still
there, for Rudolf not only disliked inventories, he also reveled
in devising complicated hiding places for his favorite tro-
phies. As after the sinking of a gigantic caravel returning from
India, astonishing debris of a once-blinding glory keeps rising
to the surface to this day.

Thus, one may hope for yet another portrait of the dodo
to emerge. While we wait, perhaps we should be grateful that
Roelandt Savery was able to resist Matthias's urgings that he

return to the Hapsburg orbit. Had he done so, his dodo paintings might have had the same fate as so much of Rudolf's collection.

---

1 Kurt Erasmus, *Roelandt Savery, sein Leben und seine Werke* (doctoral thesis), Halle-Wittenberg, Germany: Friedrichs-Universität, 1907.

2 *Ibid.*

3 Joaneath Ann Spicer-Durham, *The Drawings of Roelandt Savery* (doctoral thesis), New Haven, Connecticut: Yale University, 1986.

4 Genesis 9, 2-3.

5 Psalm 8.

6 Kemp Malone, *The Dodo and the Camel: A Fable for Children Freely Told in English by Kemp Malone, after the Danish Version of Gudmund Schütte*, Baltimore, Maryland: J. H. Furst Company, 1938.

7 Joaneath Ann Spicer-Durham, *The Drawings of Roelandt Savery* (doctoral thesis), New Haven, Connecticut: Yale University, 1986.

8 Kurt Erasmus, *Roelandt Savery, sein Leben und seine Werke* (doctoral thesis), Halle-Wittenberg, Germany: Friedrichs-Universität, 1907.

Chapter 4

Mauritius and Réunion

AT THE END OF THE SIXTEENTH CENTURY, the Dutch and the English made entrances into the Indian Ocean almost simultaneously. Both had sought and failed to find northern routes to the Indies (the British sailing west, the Dutch sailing east), and both had established East India Companies—Britain in 1600 and Holland in 1602. Both were determined to develop a far-eastern trade in competition with the Portuguese.

By 1611, the Dutch and the English were plying a new southern route to India that came to be known as "the great route." From the Cape of Good Hope, it passed south of the Mascarenes, turning north just east of Rodrigues. Although the passage through the Mozambique Channel was never completely abandoned, the Mascarenes became more attractive to Dutch and English, and eventually French, shipping. The islands provided a safe and convenient location for replenishing supplies of food and fresh water, for the occasional repair of vessels, and even for supplies of timber. Most important, they were uninhabited and therefore posed no threats of the kind that had forced the Portuguese to abandon the Muslim Comoros, where, in 1591, an entire English crew was massacred.

For four decades after the 1598 voyage of Jacob van Neck, who claimed Mauritius for Holland and brought back the first live dodos to Europe, no effort was made to settle the island. A number of vessels dropped anchor there, and accounts were written of what they encountered. The narrative of a 1607 French voyage says that the men of the party "lived on Tortoises, Dodos, Pigeons, Turtle-doves, gray Parrots and other game, which they caught by hand in the woods."[1] All visitors in this period were impressed by both the quantity of wild game available and the ease with which it could be killed.

In the log of an expedition sailing back to Holland from Indonesia, the captain refers to several feasts that his men made of dodos, apparently less disgusted than their predecessors, van Neck and his crew. He mentions that three or four—sometimes only two—dodos were enough to furnish an ample meal for all, which gives us an idea of the bird's bulk. "When Jacob van Neck was here, these birds were called Wallich-Vogels [disgusting birds], because even a long boiling would scarcely make them tender, but they remained tough and hard, with the exception of the breast and belly, which were very good."[2]

Thus the placid animals of the island were under siege long before men attempted to settle there. Even the fish, previously left alone, were pursued into the deepest pools. With their round, well-fattened rumps, dodos were obvious targets of food-seeking visitors. Sailors brought plump dodos back to the ships, whole crews making ample meals from them, with substantial leftovers. In a single hunting expedition, the men returned to their ship with 50 large birds, including 24 or 25 dodos so large and heavy that the assembled company could not eat two of them for dinner, and all the meat that remained was salted. In only three days this same crew captured another 50 birds, including some 20 dodos, all of which they brought on board and salted.

Hardly any account from this period fails to mention the enormous size of the dodo and how useful it was as preserved meat for the remainder of a voyage. Amid what was becoming a festive ongoing slaughter, only once was there a reference to the dodo trying to fight back: one sailor wrote that if the men were not careful, the birds inflicted severe wounds upon their

Vaïal soektmen Hier en vlees van't pluim gediert
Der pallembomen sap de dronten rond van staten
't Wylmen de papegai hout dat hy piept en tiert
En doet dat and're meer ook raaken inder miuten

Describing this 1648 illustration by Willem ven West Zanen, Anthonie Oudemans noted, "In the foreground can be seen two men mauling a parrot and causing it to scream, while in the background three men are busy hitting dodos with sticks while a fourth looks on, and a fifth holds up two dead dodos, their heads dragging along the ground."[3] (From Masauji Hachisuka, The Dodo and Kindred Birds; or, The Extinct Birds of the Mascarene Islands, London: H. F. & G. Witherby, Ltd., 1953.)

aggressors with their powerful beaks. The aspiring English naturalist Sir Thomas Herbert visited Mauritius in 1627, when the island was still uninhabited. His account was published in several versions, titled *Travels*, after 1643. In it, his fascination with the strange bird is clear: "The Dodo comes first. A Portuguize name it is, and has reference to her simpleness:"

> It is a bird which for shape and rareness might be called a
> Phoenix (wer't in Arabia). Her eyes be round and small, and
> bright as diamonds. Her body is round and extremely fat, her
> slow pace begets that corpulencie . . . and so great [are they in
> size] as few of them weigh less than fifty pounds: meat it is
> with some, but better to the eye than stomach; such as only a
> strong appetite can vanquish: but otherwise, through its
> oyliness it cannot chose but quickly coy and nauseate the
> stomach, being indeed more pleasurable to look than to feed
> upon. It is of a melancholy visage, as sensible of Nature's
> injury in framing so massive a body to be directed by
> complemental wings, such indeed as are unable to [hoist] her
> from the ground, serving only to rank her amongst Birds; her
> stomach is fiery, so as she can easily digest Stones and Iron.4

The last sentence refers to the presence of a stone-like formation found in the digestive tracks of a number of animals that do not have teeth. The dodo was odd in having only one of these, whereas most animals with them have more than one so that they operate in a milling fashion. But where Herbert got the idea that the dodo could digest iron is anyone's guess.

Peter Mundy's somewhat less fantastical account offers a concrete indication of the rapid decline in the dodo population. Mundy, an official of the British East India Company, was in his time one of the most wide-ranging voyagers since Marco Polo. On a visit to Mauritius in March 1634, he noted, "Dodoes, a strange kinde of a fowle, twice as big as a Goose, that can neither flye nor swymm, being Cloven footed; a wonder how it should come thither, there being none in any part of the world yet to be found."5

Four years later, returning from China, Mundy stopped at Mauritius again and went looking for the dodo. "We now met with None," he wrote in his diary. "As I remember they are as bigge bodied as great Turkeyes, covered with Downe, having little hanging wings like short sleeves, altogether unuseful to Fly withal, or any way with them to helpe themselves."[6] The formerly great numbers of dodos were now in serious decline.

François Cauche visited Mauritius in 1638 and recorded his journey in a book titled *Relations veritables et curieuses de l'Ile de Madagascar*. He claimed to have seen birds that he called *Oiseaux de Nazareth*—larger than a swan, covered with black down, with curled feathers on the rump, and similar ones in place of wings. "They are with no tongues and their scream is like a duckling's." Cauche added that their beaks were large and curved, their legs scaly, their nests made of herbs heaped together. "They make but one egg, the size of a halfpenny roll, against which they lay a white stone, the size of a hen's egg; when one kills them they have a stone in their gizzard." Even if these details accurately match many of the dodo's features, Cauche is probably confusing this bird with yet another tropical bird, the cassowary, reputed back then to have no tongue. It is possible that the designation of *Oiseaux de Nazareth* was a misunderstanding or mispronunciation of *oiseaux de la nausée*, a French term for the Dutch *Walckvogel* for dodo. Thus, a phantom species was born, the *Didus nazarenus*, which would pester ornithologists for many decades, until all scholars agreed that this variant of the dodo never existed.[7] As with Mundy, Cauche did not see any of the increasingly scarce dodos.

———————————————

At the time of Cauche's visit and Mundy's second visit, the Dutch moved to permanently occupy Mauritius, and the dodo was still going about its business, unaware of its approaching doom. The Dutch had established their first Indian Ocean settlements in Indonesia, their eyes set on pepper and other spices, especially cloves and nutmeg. In 1619, they built a city in Java that they called Batavia (later it would become Jakarta), and made it their headquarters.

Now firmly established in the Spice Islands, the Dutch set about consolidating their power in the region with the aim of forcing the English and the Portuguese out of Indonesia and its surrounding islands (an area later known as the Dutch East Indies). This strategy involved establishing reliable supporting settlements carefully dispersed along the route from Europe. To this end, in 1638, the Dutch East India Company sent an expedition to seize Mauritius and make it an official Dutch possession. At this point, the main goal of the occupation was to forestall any attempt at settlement by the Netherlands' rivals, France and England. The French, especially, had no stronghold in the Indian Ocean other than Madagascar and were known to be on the lookout for a better place to settle. The English were also becoming an increasingly dangerous threat. Beginning in 1612, they maintained a trading station at Surat on the westcoast of India north of Bombay, and were to obtain similar rights near Madras in 1639.

Since there was no one on the island to oppose them, the Dutch takeover of Mauritius was accomplished easily by a small party of twenty-five convicts, with the assistance of slaves brought from Indonesia and Madagascar. Cornelius Gooyer was named governor of the island and given a number of assignments. He was to ensure that food was available for Dutch ships calling at Mauritius, and for that purpose he was to grow crops, including tobacco, and keep cattle and poultry. Ebony trees, growing in profusion on the island, and ambergris, the secretion of whales found on the beaches, were reserved for export to Holland. It was envisioned that the island be used as a convalescence base for Dutch settlers and officials who became sick in Batavia. Unlike Java and the other islands of the Dutch East Indies, Mauritius had a healthy climate and was free from tropical diseases.

Under Gooyer, settlers raised spices, sugar, pineapples, and other tropical products, slowly but effectively undermining the natural habitat. They had brought pet dogs with them, and it wasn't long before the dogs began stealing dodo eggs from defenseless nests on the ground. The settlers also introduced goats, which competed with the dodo for food resources. More ships meant more rats coming ashore, proliferating to huge numbers and soon feasting on the eggs, even the chicks, of the slow-moving bird.

Batavia never supported the post consistently, and the settlement was aborted after 20 years. By that time, the settlers had plundered the original forests, decimating the trees to such a degree that as early as 1650, the price of ebony in Europe plummeted. Legislation was passed forbidding the felling of more than 400 trees a year on Mauritius, at that point the world's leading supplier of ebony. Meanwhile, the crops the settlers had planted failed to bear fruit—not unlike the settlers themselves, because, in an astonishing oversight, the colony had been provided with very few women.

The Dutch tried again in 1664 with an expedition sent from the Cape of Good Hope, where they had established a colony 12 years before. This time an appropriate number of prostitutes were included. But, even after this wise precaution, the new settlement did not truly flourish: by the end of the seventeenth century, there were barely 300 people on the island, including slaves, with men outnumbering women two to one.

In 1669, Commandant Fredrik Wreede, who had demonstrated ability as a leader in the military at the Cape, arrived on Mauritius to take charge. But soon after his arrival his health broke down, and he became morose, brutal, and harsh in his treatment of the settlers. By 1672, he was so hated that his subordinates drowned him. His successor was a reformed pirate named Hugo, who in his new incarnation was unable to keep discipline among the settlers. He was replaced by an autocrat named Lamotius, who managed to do some useful work but eventually became so violent that he had to be recalled in 1692.

Finally came Roelop Deodati, who ruled the island until 1703 and appeared to be effective, but the Cape colony was unable to give Mauritius the attention it needed, and swarms of fast-breeding rats kept devouring the crops. Once more the Dutch gave up and abandoned Mauritius, leaving behind sugar cane and sambar deer, both imported from Java. Some of the previously introduced cattle ran wild. Monkeys and pigs had become pests along with rats. These developments wiped out all but a few vestigial remnants of the original habitat. The dodo was not among the vestiges that survived.

Benjamin Harry, an Englishman who visited Mauritius in 1681 as a first mate on a British vessel, is credited as the last person who saw a live Mauritian dodo and wrote about it. In a manuscript now kept at the British Museum, titled *A coppey of Mr. Benj. Harry's Journal when he was cheif mate of the Shippe Birkley Castle, Captn. Wm. Talbot then Commander, on a voyage to the Coste and Bay, 1679, which voyage they wintered at the Maurrisshes*, the author recounts that on a return voyage from India, his ship was unable to pass the Cape of Good Hope, and the decision was made to "make for the Marushes."[8] Describing this landing at Mauritius, Harry followed up with observations about the island and some of its flora and fauna, including the dodo. After this brief mention, all evidence is negative. Explorer after explorer, equipped with old descriptions and vague oral accounts, sought out the island's oddest resident. But no one ever again reported seeing a dodo in Mauritius's receding forests.

Feeding the curiosity of these visitors was the fame the dodos had achieved upon being brought to Europe. By the late seventeenth century, some 14 dodos had reached Europe alive: one to Genoa, one to Germany, one to Antwerp, two to England, and five males and four females to Holland. None of them lived very long. No one seemed to have thought of joining a male and a female together in captivity (and had anyone done this experiment, it could just as easily have failed); at any rate, none of the exiled birds had descendants. By the end of the seventeenth century, the bird had become so celebrated that there were 78 different European words referring to it— *dodo, doudo, totarsen, dronte, oiseau de la nausée, walickwogel, dodo-aarsen, dodaers, dodeersen, geant*.... By then, of course, the words referred to a bird that no longer existed.

In addition to Europe, at least two dodos were taken to India as pets; their presence in Surat was noted in the accounts of Peter Mundy. Another dodo left Mauritius in a shipment destined for Japan, but there is no indication that it ever completed the journey. As much as we owe praise to Roelandt Savery for his dedication to the awkward model in the early 1600s, we have to realize that this unsolicited attention made the dodo a kind of superstar. These creatures, which had spent thousands of years in blissful peace and quiet, without a care vaguely reminiscent of this sudden tribulation, had abruptly become working birds, forced to tolerate being stared at and to pose for painters. Add the rigors of European winters, and it is no wonder that none of them survived for long.

Of the two dodos that went to England, the dead carcass of one was bought by the naturalist John Tradescant. He had it stuffed and added to his collection of specimens of unusual creatures. The catalog of this collection, published in 1656, lists "a Dodar from the Island of Mauritius. It is not able to flie being so big."[9] When Tradescant died, his collection was transferred to the Ashmolean Museum at Oxford. Thus, a badly stuffed and rather mangy dodo took up residence at Oxford in 1683.

By then it had been two years since anyone had seen a live dodo in Mauritius. The extent to which European civilization had transformed the fauna of Mauritius is more than hinted at in Jean de la Roque's account, titled *Voyage de l'Arabie Heureuse*, published in Paris in 1715.[10] The expedition's ships touched at Mauritius, then unoccupied, in 1709. In the mountains were spotted "a large number of Hogs which created a great destruction and against which a general hunt was ordained for their extermination." The pigs were so numerous that, according to de la Roque, "in one day the men killed more than 1500." Unlike other introduced species, pigs are omnivorous: not limited to a single food supply, they could prosper by preying on and eventually destroying several of the island's species simultaneously. These herds were capable of deadly, efficient predation upon the flightless dodo, or at least upon its juveniles and the nests. Calling attention to another powerful threat, de la Roque writes: "Upon walking in the island I had the pleasure to see more than 4000 monkeys in a nearby garden."[11] These animals in all probability were crab-eating macaques brought to Mauritius for their meat, but there is disagreement on whether the perpetrators were the Portuguese or the Dutch. The former openly professed their love for macaque meat, while the latter downplayed this particular gastronomic taste, but seemed to enjoy it just as much.

Commenting on de la Roque's passage, David Quammen, a modern observer and author of *Song of the Dodo*, has written that this large number of monkeys

> suggests a plague of frantic omnivores that may have made life (or at least procreation) impossible for a species of ground-nesting birds. . . . Whoever did introduce the crab-eating macaque, for whatever inscrutably stupid reason, committed a very consequential act. Those monkeys have been much underestimated . . . as a contributing factor in the extinction of the Dodo.[12]

Though we can only speculate on exactly how the dodo actually perished, we can be certain that some 50 years after the Dutch seized Mauritius, the last surviving dodo was cornered on all sides, had nowhere to go, and finally succumbed. One last person might have cooked the bird for one last formidable dinner. One last pig or monkey might have run off with the bird's last chick. Another thing is known for certain. In 1755, less than a century after acquiring it, the Ashmolean Museum at Oxford threw out the last embalmed dodo from its collection and with it, the last physical remains of the animal.

The Mauritius dodo was not the only member of the dodo family to meet extinction as a result of the discovery of its habitat by human beings. It had two relatives, cousins count-less times removed, shall we say, living on the nearest islands, Réunion and Rodrigues. In all likelihood, the evolutionary path of this family started with a single prototype landing on all three Mascarene Islands. And since there was no contact between the three widely separated specks of dry land, the original dodos diverged over thousands of years into three related, yet distinct groups. Discovered by Europeans at roughly the same time, all three birds were exterminated in an equally speedy fashion.

The Mauritius dodo was dark in color. On Réunion, the dodos were yellowish-white, with black-tipped wings and other distinguishing features. No one paid much attention to this white dodo until after it was extinct.

For one thing, the island of Réunion is larger and more mountainous than Mauritius, making it less attractive for set-tlement. It still went by the Portuguese name of Mascarenhas when it was visited in 1614 by the English captain Samuel Castleton during a voyage to Oriman, India. Twelve years later, an account of Castleton's voyage was published, written by Castleton's pilot, John Tatton. On Réunion, according to Tatton, they encountered

> a great fowl of the bigness of a Turkie, very fat, and so short-winged that they cannot fly, being white, and in a manner tame....
>
> This island that the Portuguese called Mascarenhas, and the French now call Bourbon, was then desert, but it was full of birds of land of all species, pigeons, big parrots, and one other sort of bird as big as a goose, very fat, with short wings that don't allow it to fly. It has been called after the giant, and Mauritius also produces many of these. It is white, and so naturally limited that he let men grab it with their hands, or, at least, being little scared at the sight of the sailors, who could easily kill several with blows from sticks and stones. In general these birds are in such abundance in these islands that ten sailors can amass in one day enough to feed fourty.[13]

Going farther inland, the Englishmen found a large stream covered with geese and ducks. In these waters, there were also eels "that many consider of the best taste in the world." Admiring their size, Tatton found they weighed 25 pounds apiece. "If they are hit by a blow with the knife, they swim for two or three brasses, but then they stop and they let the men easily grab them." The author repeats with pleasure that this is the best fish he ever tasted. All in all, he concluded, Réunion was "an admirable place for the refreshment of voyagers."[14]

In 1619, the Dutch traveler Willem Bontekoe van Hoorn spent three weeks on Réunion and described a "*dadeersen*" similar to that encountered by Castleton's crew. The journal of this voyage contains the best description of the white dodo that has come down to us, and it became a favorite of the well-traveled and influential French patron of natural sciences, Melchisedec Thevenot, who was Louis XIV's librarian. As a result of Thevenot's approval, it was hailed in its time as much more reliable than other accounts of the same sort, and its descriptions of wildlife are unusually lavish.

"*Pes et Caput uni Reddentur formae.*"
("The foot and the head acquire one single shape.")
—Horace

This copper engraving by Willem Bontekoe, first published in van den Broeke's Journal, *was probably drawn from life, although the beak is poorly executed.* (From Hugh Edwin Strickland, *The Dodo and Its Kindred; or, The History, Affinities, and Osteology of the Dodo, Solitaire, and Other Extinct Birds of the Islands Mauritius, Rodriguez, and Bourbon*, London: Reeve, Benham, and Reeve, 1848. From the collections of the Ernst Mayr Library of the Museum of Comparative Zoology, Harvard University.)

Bontekoe left the Dutch port of Texel with three ships in 1618, sailing to India on a trading mission. In a typical journey of the time, Bontekoe's ships were lashed by storms, beset by illness, and becalmed for weeks at a time. Bontekoe's own ship, *Nouvelle Hoorn*, ended up alone off the coast of Madagascar. At this point, with severe illness among his crew, Bontekoe changed course for Mauritius, but arrived at Réunion instead. Shallow waters and hidden reefs kept the ship offshore, but "those sick [and] burning to go ashore" went via raft to the island. They found a multitude of turtles there. Bontekoe landed most of the remainder of his crew, and on shore, the condition of the most ill of his men seemed to improve. According to the journal, "It was quite a touching spectacle to see them arrive over the grass, & roll over it as in a place of delights. They assured the others that this situation alone was already relieving their pains."[15]

The bounty of the place was impressive. They found pigeons that "let themselves be killed with bare hands or by throwing sticks, without a single movement of defense. Of these, they took, from the first day, more than two hundred. The turtles were not any harder to seize." The following morning, Bontekoe found

> a good bay, with sand in the bottom. At a little distance
> inland, there was a lake where the water was not completely
> fresh. Bontekoe saw many geese, pigeons, gray parrots,
> & other birds. He found up to twenty-five turtles, in the
> shade, under one single tree. The geese never took flight and
> let men kill them without leaving their places. They were so
> fat that they could hardly walk. If they took a parrot or any
> other bird, and tortured it until it started screaming, those of
> the bird's species would come around and fly close to the
> victim, as though to protect it, and therefore were very easy
> to seize.[16]

After scouting the entire bay, Bontekoe brought the news of his discovery to the sick, who happily got on board again in the hope of finding an even more comfortable retreat. "All people in the crew had permission to come to shore and look for fresh water in the woods. Eight men entered a lake with poles and ropes and took plenty of fish, including *carps*, soles, and a kind of fat salmon with very good taste. . . ."

And then, "They also saw Drontes, whom the Dutch also name dod-aers, a kind of bird that has small wings and whose fat makes them very heavy." A great feast began:

> They made sticks with wood that were very convenient to roast the birds; & by greasing them with turtle oil they rendered them as delicate as though they had been prepared with all the amenities of a great kitchen. They discovered another river with quite beautiful water, which was full of packs of eels. They took their shirts and held them by the tips, thus using them as nets and capturing amazing numbers of these eels, whom they found to be quite tasty. There were also some goats but these were quite wild, and their horns being half-eaten by worms no one wanted to taste them.[17]

Nevertheless, the native fauna was already learning the hitherto unknown meaning of fear and self-defense. "Within the space of around twenty days, the birds of the island, rendered wild by this continuous hunt, were now taking flight as soon as a man came near them."[18]

After amassing a large provision of salted fowl and turtles, Bontekoe and his men finally left the bountiful island that restored their health. Again they headed for Mauritius but soon had regrets for having left Réunion. Their ship was in no shape to cover the distance; a fire on board seemed at first to kill everyone except for Bontekoe and one sailor, but then others survivors appeared. They built a raft, made a sail out of their shirts, and proceeded east with nothing more to eat than eight pounds of biscuits. After these were gone, they caught flying fish and ate them raw, and they discussed eating one another, starting with the youngest on board.

When, after 13 days, they finally reached land, it was the island of Sumatra in western Indonesia, thousands of miles from Réunion. Bontekoe survived and went on to join other voyages.

After Castleton and Bontekoe, no one else ever wrote about the white dodo while it was still alive.

―――――――――

The French were the last European explorers to arrive at the scene. The English and the Portuguese were already firmly established in India, and the Dutch ruled over Indonesia and the Cape. The newcomers were left with little more than Madagascar.

In 1638, the year the Dutch attempted the first permanent settlement on Mauritius, the French took possession of Rodrigues and Réunion. They named the latter Bourbon after their royal family. The unrealized French ambition to establish a colony in South Africa accounted for several naval skirmishes with the Netherlands and led to a vigorous effort to settle Réunion. Given the nature of the impulse behind settlement, Réunion was the only place in the Mascarenes where matters were taken seriously enough to include enough females, accounting for a much faster population growth than at Mauritius during the same period. In addition, plantations were established that produced good crops of coffee beans. The attack on Fort Dauphin, the French base on Madagascar, by the indigenous Madagascans added impetus to the settlement of Réunion and gave rise to the ambition to take over nearby Mauritius, where the dodo still agonized.

On Réunion, the settlers wasted no time in hunting the local wildlife into extinction. A couple of white dodos were shipped to Europe, one about 1640 and the other around 1685, where some scholars believe the birds were painted by Dutch artists.

And then they vanished in silence. In 1801, the French government commissioned a survey of the people, resources, flora, and fauna on Réunion. Finding no evidence of the white dodo and, inferring from the dates of extinction for the dodo of Mauritius and the solitaire of Rodrigues (all by negative evidence), they concluded that the crazy bird had most likely become extinct more than a century earlier.

———

1 Hugh Edwin Strickland, *The Dodo and Its Kindred; or, The History, Affinities, and Osteology of the Dodo, Solitaire, and Other Extinct Birds of the Islands Mauritius, Rodriguez, and Bourbon*, London: Reeve, Benham, and Reeve, 1848.
2 *Ibid.*
3 Anthonie C. Oudemans, "On the Dodo," *Ibis* 10 (1918), n. 6: 316.
4 Hugh Edwin Strickland, *The Dodo and Its Kindred; or, The History, Affinities, and Osteology of the Dodo, Solitaire, and Other Extinct Birds of the Islands Mauritius, Rodriguez, and Bourbon*, London: Reeve, Benham, and Reeve, 1848.
5 Peter Mundy, *The Travels of Peter Mundy*, London: The Hakluyt Society, 1919.
6 *Ibid.*
7 Abbé Prévost, *L'Histoire générale des voyages*, Paris: 1750.
8 As quoted in: Hugh Edwin Strickland, *The Dodo and Its Kindred; or, The History, Affinities, and Osteology of the Dodo, Solitaire, and Other Extinct Birds of the Islands Mauritius, Rodriguez, and Bourbon*, London: Reeve, Benham, and Reeve, 1848.
9 As quoted in: Robert Silverberg, *The Auk, the Dodo, and the Oryx: Vanished and Vanishing Creatures*, New York: Thomas Y. Crowell Company, 1967.
10 Jean de la Roque, *Voyage de l'Arabir heureuse: Par l'Ocean oriental, & le detroit de la mer Rouge, fait par les françois pour la première fois, dans les années 1708, 1709 & 1709; avec la relation particulière d'un voyage fait du port de Moka à la cour du roi d'Yemen, dans la seconde expédition des années 1711, 1712 & 1713; un mémoire concernant l'arbre & le fruit du café, dressé sur les observations de ceux qui ont fait ce dernier voyage; et un traité historique de l'origine & du progrés du café, tant dans l'Asie que dans l'Europe; de son introduction en France, & de l'établissement de son usage à Paris*, Paris: A. Cailleau, 1716.

11 *Ibid.*

12 David Quammen, *The Song of the Dodo*, New York:
    Simon & Schuster, 1996, p. 270.

13 As quoted in: Abbé Prévost, *L'Histoire générale des voy-
    ages*, Paris, 1750.

14 *Ibid.*

15 Willem Ysbrandsz Bontekoe and C. B. Bodde-
    Hodgkinson & Pieter Geyl (trans.), *Memorable
    Description of the East Indian Voyage, 1618–25*, London:
    G. Routledge & Sons, 1929.

16 *Ibid.*

17 *Ibid.*

18 *Ibid.*

Chapter 5

Rodrigues

As the smallest of the Mascarenes, only ten miles long and four miles wide, and the most isolated, Rodrigues attracted few ships and was never the target of a serious attempt at settlement until the end of the seventeenth century. This was nearly two centuries after its discovery by the Portuguese and half a century after the establishment of a more or less permanent presence of men on Mauritius and Réunion. For this reason, it was the last stronghold of the dodo family. The unusual nature of the first settlement of Rodrigues by a party of no more than eight men, and the character of their leader, a French farmer named François Leguat, gave rise to the most extensive and dependable description of the dodos.

The detailed narrative by Leguat of his two-year stay on Rodrigues, written and published by him in London in 1708, has survived the critical scrutiny of generations of experts to become the definitive account of the fauna and flora of the island before the full onslaught of the European invasion. But for this devout, gentle, and dedicated French Robinson Crusoe, we wouldn't be able to guess, with all the tools now at hand for analyzing the past, how the dodo reproduced, or what its feeding habits were. Likewise, how the entire ecosystem that made this bird first so viable (and then so utterly nonviable) worked before the arrival of human settlers would also be a mystery. The story of Leguat's voyage warrants detailed treatment, including samples of his remarkable descriptions in his own words.

Little is known of François Leguat's early years, beyond the fact that he was the son of one Pierre Leguat and he was born in the small east-central French province of Bresse, five or six years before the deaths of Cardinal Richelieu and King Louis XIII in 1642 and 1643. He was a Huguenot, that is to say, a French Protestant, although whether by birth or conversion is unknown. Quite likely he was a follower of some of the teachings of the sixteenth-century Swiss reformer John Calvin.

We do know that Leguat was over 50 years of age when he was driven into exile as a consequence of Louis XIV's revocation of the Edict of Nantes in 1685, which made it a crime to be a Protestant in France. Many Huguenots who did not renounce their heretical beliefs were massacred, while others fled.

François Leguat was one of a group of exiles who reached the Netherlands in 1689. Like the English Pilgrims settled Massachusetts earlier in the seventeenth century, the Huguenots sought Dutch assistance in finding a place where they could live in peace. Meanwhile, the Netherlands and England, stung by Louis XIV's military ambitions in the Low Countries as well as his blatant attempts to foment rebellion in Britain and thereby restore the crown to the Catholic House of Stuart, entered an alliance with Austria and several German states against France. This conflict, which came to be called the War of the League of Augsburg, would drag on indecisively for eight years.

Not long after their arrival in the Netherlands, Leguat and his friends learned of a project to establish a colony of French Protestant refugees on a Dutch-controlled island in the Indian Ocean. The project was the brainchild of the Marquis Henri du Quesnes, son of a celebrated French naval commander, the latter a Protestant whose exploits had made him a favorite of Louis XIV and thereby exempted him from the revocation of the Edict of Nantes.[1]

VOYAGE
ET
AVANTURES
DE
FRANCOIS LEGUAT,

& de ses Compagnons,

EN DEUX ISLES DÉSERTES
DES
INDES ORIENTALES.

Avec la Rélation des choses les plus remarquables
qu'ils ont observées dans l'Ifle MAURICE, à BATA-
VIA, au Cap de BONNE-ESPERANCE, dans l'Ifle St.
HELENE, & en d'autres endroits de leur Route.

Le tout enrichi de Cartes & de Figures.

TOME PREMIER.

A LONDRES,

Chez DAVID MORTIER, Marchand Libraire.

MDCCVIII.

The title page of François Leguat's Voyage, *a volume that chronicles both a failed attempt to colonize Rodrigues and, in the process, the discovery of the life and habits of the solitaire, the long-necked cousin of the dodo that lived there.* (François Leguat, *Voyage et avantures*, 1708. Courtesy of the General Research Division, The New York Public Library, Astor, Lenox, and Tilden Foundation.)

Under the auspices of the Dutch States-General and the directors of the Dutch East India Company, Henri du Quesnes organized an expedition to establish of a colony of French Protestant refugees on Réunion, then known by the French name of Bourbon. Du Quesnes chose the island on the presumption that the French had pulled out, leaving the island up for grabs. He rechristened the island with the much more suggestive name of "Eden" in a detailed prospectus that he wrote and published, which drew upon previous glowing accounts. (There is no evidence that du Quesnes ever visited the island himself.) The prospectus was designed to entice colonists who, like Leguat, had not fled poverty or criminal pasts but worked hard and even been men of means, but now were without family or other ties to keep them in Europe. It was not long before du Quesnes had chartered and began fitting out two ships to carry a large party of Huguenots who had signed on for a free passage to Eden.

According to some scholars, du Quesnes chose to organize this colony in order to go down in history as the man who had conceived, overseen, and made possible with his patronage the utopian dream of a gentle, vice-free civilization started anew on a desert island. If he had more prosaic thoughts about making profits through, say, having a percentage of the goods shipped in the future from Utopia, he kept those thoughts to himself.

Leguat made such a good impression on the organizers that he was named the major of one of the two ships. By occupying this post, Leguat was charged with keeping his shipmates orderly, organized, and filled with unyielding faith.

It is not hard to imagine how excited the Huguenots were by the possibility of building their own colonial paradise, even if it was on a faraway dot in an unknown ocean. Finally, they would escape religious prosecution and live in peace by their own rules, achieving something that had eluded them until now. The obstacles posed by a long and dangerous sea voyage were overcome by the marvelous descriptions of Réunion that had come back to Europe, and which du Quesnes had woven into his prospectus. Leguat seems to have swallowed these wonderful accounts whole, because he incorporated them into his own narrative.

By all existing descriptions, du Quesnes's island was a marvelous place, a terrestrial paradise with healthy air, to judge from the number of sick people who had gone ashore and promptly recovered their health. "The sky is clear; the Exhalations of the Earth, as well as those of the Aromatick Plants and Flowers . . . perfume the Air, and they breath in a balmy Spirit, equally agreeable and wholesome." Abundance was a major theme, especially of springs and rivers gushing with clear and wholesome water, full of fish. The beaches abounded with shells, coral, ambergris, and turtles with delicate flesh and excellent oil. Absence of malice was another theme on the island. "There is no venomous Creatur . . . neither in the water, neither on the dry Land; whereas almost all other hot Countries are full of Snakes, and such sort of Animals, whose Sting or Bite is dangerous, if not mortal. The same thing is affirmed of the Plants and Fruits here." An Eden without serpents or dangerous fruit—what more could a good Christian ask for? The forests contained cedar, ebony, and palm, and the excellent soil yielded a score or more of citrus and fruit trees and hundred of edible plants.[2]

In the highly unlikely event, according to du Quesnes, that farming would fail, settlers could live simply by feasting on the native birds. In his listing of birds he expected to encounter, Leguat mentions for the first time what probably was the white dodo of Réunion. Leguat's first reference to these birds, based only on hearsay, appears in the original French version of his 1708 account as "géants," or giants. In later English translations these giants became peacocks, a bird that was certainly never native to the Mascarenes—but, as far as European readers were concerned, peacocks were an essential feature of a Country of Delights.

"The Fowl there [that] are most plenty [plentiful] on this
Island," wrote Leguat, "are Partridges, Doves, Ducks, Wood-
Pigeons, Woodcocks, Quails, Black-Birds, Puets, Trushees,
Geese, Coots, Bitterns, Parrots, Herons, Fools, Frigats,
Sparrows, Peacocks, and abundance of other small Birds."3
All of them would make for excellent meals, according to the
reports, and there was even an additional source of nutrition:
"There are Bats whose Bodies are bigger than a Hen's, and
the Flesh of them very pleasant to eat, when a Man gets over
that Aversion to them which is begot by Prejudice."

On the downside, "little Sparrows, which, like Flowers and
Butterflies, seem to have been made only to embellish Nature,
multiply so fast, that, to say the truth, they are troublesome. They
come in Clouds, and carry away the Corn that is sown, if great
Care is not taken of it; which is doubtless an Inconvenience; but a
little Gunpowder soon frightens them away."

Caterpillars and flies could also be "a little vexatious,"
according to Leguat's sources, and there was also the danger
of seasonal hurricanes. But even a Huguenot had to admit that
no paradise set on earth could be absolutely perfect. On bal-
ance, du Quesnes had carefully analyzed the potential of
Réunion for offering the settlers a happy and harmonious
living. And there seemed to be little reason to doubt his over-
all finding: "It is certain the Isle of Eden is of sufficient Extent,
to contain easily a long descent of Generations, of whatever
Colony will settle there."4

One very serious fly in the ointment must have been du
Quesnes's sudden announcement that no women would be
taken on the voyage. Du Quesnes decided this after receiving
the news that the French, far from abandoning Réunion, had
sent a fleet of seven warships to the region. To avoid a confrontation with the French, the two Dutch ships would be
sent out disarmed (i. e., without cannons). This meant that
the crew and passengers should be able to defend themselves
individually, which excluded women who could not be
expected to do so. An unprovoked attack by the French on a

boat carrying unarmed religious refugees was unlikely, however, because there were certain unwritten rules of honor functioning among European nations about such situations. The Huguenots were told that women could join them after they established themselves on the island.

Next, du Quesnes learned that the French had re-annexed Réunion to the French East India Company in 1674, 15 years before, and the French fleet had been sent to back up this claim. This meant there was no way a colony of Huguenots could be established.

Instead of sharing this information with Leguat and the others, du Quesnes scaled back the project from two ships to a small frigate, *La Hirondelle* (Swallow). The commander, a certain Monsieur Valleau, was directed to reconnoiter the islands of the Mascarene group and take possession of whatever island was found unoccupied and suitable for colonization. This radical change of plans was never communicated to the small band of heretic adventurers who embarked as emigrants under the idea that they were to be landed on Réunion, the island named Eden, their terrestrial paradise.

By the time the *Hirondelle* was ready to sail, a number of prospective passengers had dropped out of the risky trip to Eden. With their numbers now down to ten, Leguat and his followers resigned themselves to the hands of Providence and departed from Amsterdam on July 10, 1690.

From the beginning, the journey was not a happy one, but Leguat took from it all the good things a man possibly could, including detailed accounts and meditations concerning the maritime species and natural phenomena observed at sea. His Huguenot faith was unshaken; his tone was at times so pious and sincerely confident in the care of God that it is hard to believe such passages could come from the pen of anyone in such dire circumstances. For instance, of a narrow escape from a shipwreck and a pursuit by a French privateer, Leguat wrote: "We were all of us convinced by this Double Deliverance the same day, that we had been under the singular Protection of the Almighty, and we rendered the Thanks that were due to his Divine Favour."5

Leguat chronicled in unrelenting detail all the manifestations of life the passengers could observe while sailing. These included porpoises, whales, flying fish and other fish, and numerous kinds of sea birds. Characteristic of Leguat's writing is the interspersing of the voyage account with lengthy meditations on God, religion, and the corruption of Christian worship by various churches and those who presided over them. An excellent example is his treatment of the ritual that marked the ship's crossing the equator on November 23, 1690. To his dismay, they were "obliged to undergo the impertinent ceremony of Baptism, at least all those who had not assisted at the same Festival before, or would not buy themselves off for a piece of Money. It is an old costume, and will not be abolished without difficulty." A long, heartfelt description of the "Festival" followed:

> One of the Seamen who had past the Line before, dressed himself in rags, with a beard and Hair of Hards of Hemp, and blacked his face Soot and Oil mixed together. Thus Equipped, holding a Sea-Chart in one Hand, and a Cutglass in the other, with a Pot full of Blacking Stuff standing by him, he presented himself upon Deck attended by his Suffragans, dressed as whimsically as himself, and armed with Grid-Irons, Stoves, Kettles, and little Bells; with which rare Instruments they made a sort of Music, the goodness of which may be easily imagined. They called those that were to be initiated into these Rites and Mysteries one after another, and having made them sit down on the edge of a Tub full of Water, they obliged them to put one Hand on the Chart, and promise that on the like Occasion they would do to others what was at that time done to them. Then they gave them a mark in the Forehead with the stuff out of the Pot, wetted their faces with Sea Water, and asked them if they would give the Crew any thing to drink, promising them they would in such case let them go without doing any further Penance. As

for those who paid nothing they were thrown into the Tub of Water over Head and Ears, and then washed and scrubbed every where with the Ships Ballast; and I believe this scrubbing and washing lasted much longer than those who were so treated desired. Every Nation practices this ridiculous Custom after a different Manner.[6]

It is not clear exactly when, but after months at sea, the *Hirondelle* reached the Dutch colony at the Cape of Good Hope, where it put in for three weeks of rest and reprovisioning. During the stopover, the Huguenots were beset by conflicting rumors about their destination. According to one, a French naval fleet had landed 300 men on "Eden." Another said the fleet had never reached the island, which was inhabited by only a few families. But nothing they were told contradicted their vision of the island as a paradise of beauty and fertility. They therefore resolved to press on to Dutch-controlled Mauritius for further information. A storm blew them off course, and after some aimless drifting, they came upon not Mauritius, but Réunion itself. Fighting difficult winds, the ship came close enough to shore for Leguat and the others to see there was no sign of a French naval presence and the island, much to their joy, seemed to deliver on du Quesnes's promises:

> We flattered ourselves with Hopes that it was the Isle of Eden; and we made merry with the thoughts of setting foot on the Land we so much desired as designed for the place of our Habitation. We discovered several Beauties in this admirable Country, from the place where we stopped to view it: All that part of it which presented it self to our View appeared to be a Level, with Mountains rising in the Middle; and we could easily discern the agreeable mixture of Woods, Rivers and Valleys enameled with a charming Verdure. If our Sight was perfectly well pleased, our Smell was no less; for the Air was perfumed with a Delicious Odour that ascended from the Isle, and that pleasantly arose from the abundance of Limons and Oranges which grow there. This sweet Odour struck us all alike, when we came at a certain distance from the Island: some agreeably complained, that the Perfume

hindered them from sleeping, others said that they were so
embalmed with it, that it was much a Refreshment to them, as
if they had been fifteen days a-shore.[7]

At this point, the *Hirondelle*'s captain dashed his passengers' hopes by revealing he had orders not to land them on Réunion and had done his best to avoid it altogether. He promised to take them to another island every bit as good. Weakened from scurvy, the Huguenots were in no position to argue. After a month of sailing against the wind, they covered the 450 miles that separated Réunion from Rodrigues and reaching the uninhabited island on May 1, 1691. Again, the Huguenots were pleased by what they saw:

> The Island afar off, and near at hand, appeared to us very
> lovely . . . and indeed this little new World seemed full of
> Delights and Charms. The Face of it was extremely Fair. We
> could hardly take our Eyes off from the little Mountains, of
> which it almost entirely consists, they are so richly spread
> with great and tall Trees. The Rivers that we saw run from
> them watered Valleys, whose Fertility we could not doubt of;
> and, after having run through a beautiful Level, they fell into
> the sea, even before our Eyes. . . . We admired the secret and
> wonderful Ways of Providence, which, after having permitted
> us to be ruined at home, had brought us thence by many
> Miracles, and now dried up all our tears, by the sight of the
> Earthly Paradise it presented to our view; where, if we would,
> we might be rich, free and happy; if contemning vain Riches,
> we would employ the peaceable Life that was offered to us, to
> glorify God and save our Souls.[8]

After two weeks, the *Hirondelle* set sail again, leaving Leguat and seven others on Rodrigues. Choosing a site not far from the sea and next to a brook, "the water of which is clear and good," they set about building sleeping huts from plantain trunks, which they roofed with plantain leaves, and a central gathering place for cooking and dining. They cleared land for planting and enclosed their plots to keep out the massive turtles, themselves an excellent food source. They had mixed

Leguat's party of Huguenots trying to establish utopia on Rodrigues. A total of seven huts were built to house eight men. The settlement also had a modest town hall where the pious settlers had their meals, debated issues, and joined in prayer nearly every day. (The John Carter Brown Library at Brown University.)

results from the seeds they had brought with them—their wheat failed so they had to do without bread, but were able to grow fruit and melons that were enormous and delicious. In fact, food was no problem, since the island abounded with game in the form of turtles and birds, which could be knocked down with sticks or stones. It was by no means a hardscrabble existence, which meant they had much spare time for walking, talking, praying, and thinking.

Leguat himself spent his spare time making thorough observations of the island's natural features, and nothing he encountered was as remarkable as the bird, which he dubbed "solitaire" because "it is very seldom seen in Company, though there are abundance of them." Discoursing on certain habits of the solitaire and other native birds, Leguat makes interesting observations about the factors that conspire to doom out-of-the-way island populations. Concerning what he called "wood-hens," he noted that their fat made them too heavy to fly and they displayed the islander insouciance that made them such easy prey. Holding a red object in your hand would rile a wood-hen so much "that it would leap at you to pluck the object from your hand, providing the opportunity to capture them with ease."9

But it was the solitaire—the cousin of the dodos of Mauritius and Réunion—that truly fascinated Leguat. From his account emerges an extraordinary portrait, not of a foolish or crazy bird, but of an erect, dignified, even chivalric creature that could teach human beings a thing or two when it came to good behavior and deportment.

Leguat described the male solitaire as having feathers that were brownish-gray, with feet and beak like a turkey's, "but a little more crooked. They have scarce any Tail, but their Hindpart cover'd with Feathers is Rounds, like the Crupper of a Horse, they are taller than Turkeys. Their Neck is straight, and a little longer in proportion than a Turkey's, when it lifts up his Head. Its Eye is black and lively, and its Head without Comb."10

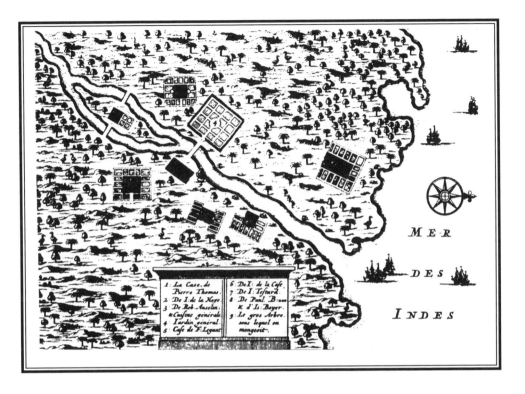

The "village" was surrounded by several patches of land dedicated to agriculture. With the exception of a few varieties of melon, the Huguenots' farming efforts proved to be a major failure. Thus restrained from developing what they had hoped would be one of their major activities and sources of well-being, Leguat's men took to fishing and hunting, while their leader took it upon himself to chronicle as many aspects of Rodrigues's natural history as possible. (François Leguat, *Voyage et avantures*, 1708. Courtesy of the General Research Division, The New York Public Library, Astor, Lenox, and Tilden Foundation.)

The solitaire from Rodrigues as Leguat saw it during the two-year exile of the French Huguenots. From their first encounter, the long-necked walking bird fascinated the French author with its elegant looks and gracious demeanor. Inspired by the animal's breeding habits, Leguat wrote some interesting paragraphs in which he drew analogies between the "marriage" of the solitaires and human relationships. (The John Carter Brown Library at Brown University.)

The solitaire never flew because its wings were too small to support the weight of its body; the birds used their wings, such as they were, only

> to beat themselves, and flutter when they call one another. They will whirl about for twenty or thirty times together on the same side, during the space of four or five Minutes: the Motions of their Wings makes them a noise very much like that of a Rattle; and one may hear it two hundred Paces off. The bone of their wings grows greater towards the Extremity, and forms a little round Mass under the Feathers, as big as a Musket Ball: that and its Beak are the chief Defense of these Birds. . . .[11]

Not surprisingly, the birds were a relatively easy prey, especially in open spaces, "because we run faster than they do, and sometimes we approach them without much Trouble. From March to September they are extremely fat, and taste admirably well. Especially while they are young, some of the Males weigh forty five pound."[12] As François Cauche described in his account of visiting Mauritius in 1638, Leguat noted the presence of:

> . . . in the Gizzards of both Male and Female a brown Stone, of the bigness of a Hens' Egg: it is somewhat rough, flat on one side, and round in the other, heavy and hard. We believe this Stone was there when they were hatched, for let them be never so young, you met with it always. They have never but one of them, and besides, the passage from the Craw [the crop of fowls] to the Gizzard is so narrow, that a like Mass of half the bigness couldn't pass. It served to wet our knives, better then any other Stone whatsoever. . . .[13]

As for the females, Leguat found them nothing less than beautiful. "They walk with so much pride and grace altogether that we cannot help admiring and loving them; in such a way that often their fine features have saved their lives." In color, some were "fair," others brown. Not one feather was out of place anywhere on their bodies, because they carefully groomed themselves with their beaks. "The Feathers on their Thighs are round like shells at the end, and being there very thick, have an agreeable effect: they have two Rising on their Craws, and the Feathers are whiter there than the rest, which lively represents the Neck of a Beautiful Woman."[14]

When it was time to nest, the solitaires chose "a clean Place," gathering palm leaves and heaping them a foot and a half off the ground. The female laid only one egg at a time, which was much larger than a goose's egg. The female and her male partner took turns sitting on the egg until it hatched, some seven weeks after laying. During this period and for several months afterward, during which the chick depended on its parents for feeding, the solitaire couple would not tolerate any other solitaire coming within 200 yards of the nest.

In defending a nest, a male solitaire never drove away female intruders; instead, he summoned his partner with a noise of his wings, and she drove off the intruding female. Likewise, the male partner was responsible for driving off male intruders. Even after raising a young bird and letting it go out on its own, the solitaire couple, Leguat noted with satisfaction, remained

> always together, which the other Birds are not, and though they happen to mingle with other Birds of the same Species, these two Companions never disunite. We have often remarked, that some days after the young one leaves the Nest, a Company of thirty or Forty Birds brings another one to it; and the new fledged Bird with its Father and Mother joining with the Band, march to some by Place. We frequently followed them, and found that afterwards the old ones went each their way alone, or in Couples, and left the two young Ones together, which we called a Marriage.[15]

No doubt aware of how readers will react to this passage, Leguat goes even further, detecting in the society of solitaires practices human beings would do well to adopt:

This Particularity has something in it that looks a little
Fabulous, nevertheless, what I say is sincere Truth, and what
I have more than once observed with Care and Pleasure:
neither could I forbear to entertain my mind with several
Reflections on this Occasion. I send mankind to learn of the
Beasts. I commended my Solitaries for marrying young (a
piece of wisdom practiced by our Jews) for satisfying Nature
in a proper time, and comfortable with the intention of the
Creator. I admired the Happiness of these innocent and
faithful Pairs, who lived so peaceably in constant Love: I said
to myself, if our Pride and Extravagance were restrained, if
Men were or had been as wise as these Birds, to say all at
once, they would marry as these Birds do, without any Pomp
or Ceremony, without Contracts or Jointures, without
Portions or Settlements, without Mine or Yours, without
subjection to any Laws, without any Offence, with which
would be most pleased, and the Common-Wealth most
benefited; for Divine and Human Laws, are only precautions
against the disorders of Mankind.[16]

Be that as it may, since Leguat and his comrades were all
males, on Rodrigues they were in a crucial sense more solitary
than the solitaire because they had no mates to marry with
pomp and ceremony or without. Despite the island's bounty
and peace, something was clearly missing—that which hap-
pens to make human settlements viable. In one passage
Leguat likened their village to the Seven Hills of Rome: "Had
there been Women amongst us, 100 years hence, instead of
seven Hutts, one might have reckoned seven Parishes." The
Huguenots coped to the best of their abilities to life on an isolated rock, without prospects of sex or family life. They
prayed, they worshipped, and Leguat sought refuge in detail-
ing all the minutiae of Rodrigues's natural history.

After two years, realizing that they had been abandoned, if not
forgotten, by the rest of the world and faced nothing short of
extinction, they built a raft and set off for Mauritius 360 miles
away. They managed to make it, but at Mauritius, then no

more than a Dutch convict establishment, the cruel governor had them imprisoned as enemy aliens on an exposed rocky islet far offshore, where one of their number died during an escape attempt. But they had contrived to have news of their plight sent to Europe, and as a result they were sent on to Batavia in Indonesia, still in confinement. Arriving there in December 1696, they spent more time in jail, until an examination by the Dutch Council established their innocence. It was not until March 1698, following the proclamation of the Peace of Ryswick in late 1697, which settled the protracted War of the League of Augsburg, that Leguat and two others, the sole survivors of the original Huguenot party, landed at the Dutch port of Vlissingen (Flushing).

A key provision of the Peace of Ryswyck was that France now acknowledged the legitimacy of the passing of the British throne to the Protestant William of Orange and would no longer aid Stuart attempts to regain it. As a result, crowds of French Protestant refugees streamed into Britain, where they received a warm welcome. Then around 60 years old, Leguat joined this migration and appears to have remained in England for the rest of his life. It was when he was in his seventies that his book *Voyages et Avantures de François Leguat, & de ses Compagnons, en deux isles désertes des Indes Orientales: Avec la Rélation des choses les plus remarquables qu'ils ont observées dans l'Isle Maurice, à Batavia, au Cap de Bon Esperance, dans l'Isle St. Hélène, & en d'autres endroits de leur Route: Le tout enriche de Cartes & de Figures* was published in London, in French and English simultaneously, while another French edition was published in Amsterdam, and a Dutch version was published in Utrecht. A German translation followed in 1709. Another French edition was produced in London in 1720, and an abridged edition appeared as late as 1792. There was also an abridged translation published under the title *The French Robinson*, and yet another one was prepared but never published in 1846.

Well-received and favorably reviewed from the start, it propelled Leguat from miserable refugee to polished socialite hobnobbing with such scientific luminaries of his age as Baron Albrecht von Haller, the Calvinist founding father of physiology. Leguat died in London in September 1735 at about the age of 96. The word "Providence" appears in the first and last pages of his book, and countless times in between. His faith in the Bible and reliance on divine support, both springing from Huguenot conviction, upheld his equanimity during long years of increasingly trying exile. He returned from these ordeals a calm and gentle old man, and proceeded to become a famous writer and then die at a very old age in security and comfort, but still in exile. His bones were never returned to France.

And the solitaire? A key to its fate was in its breeding habits as described by Leguat. Only one egg was laid at a time, followed by an incubation period of two months, plus many more months of parental care. The solitaire was doomed because it was a slow breeder; it could never make up its loss in numbers to hungry settlers and their predatory animals.

The history of Rodrigues after the Huguenots departed on their flimsy raft is spotty. It appears that between 1706 and 1707, some English officers stayed for awhile and surveyed Port Mathurin, where Leguat and the Huguenots had built their huts. Evidently the island remained a French possession, at least in the eyes of the French. In 1712, the Minister of Marine in France requested information on the capabilities of the island for providing shelter to anchoring boats and furnishing goods for Mauritius, a newly-acquired and very important stopping point en route to India. A report was issued describing Rodrigues as difficult for ships to enter, but able to provide for the anchorage of vessels of 30 guns.

Furthermore, it was stated, apart from the quantity of tortoises available, the island was of no use for the French East India Company. Nevertheless, the French authorities did not give up this useless possession. In 1725, a superintendent and a guard were posted on the island, and in 1740 the post was entrusted to "a Negro family."[17]

At some time previous to 1756, a regular French establishment was formed for the preservation of turtles, destined to supply Mauritius and Réunion with fresh meat. "This fishery [the sea-turtles]," wrote French chronicler Charles Noble in 1756, "is thought so useful at Mauritius, that they have always a sergeant's party on the little island of Rodrigues, who collect all the fish they can, for the boats that are sent to bring them at certain times, and the ships that generally touched there on their way to Mauritius. There is also a particular spot of ground enclosed here, for keeping and breeding land tortoises for the same purpose."[18] Such traffic was sufficient to doom the solitaire.

In May 1761, during the Seven Years' War, once again pitting the French and their allies against the British and their allies (this war is better known in the U.S. for its North American extension called the French and Indian War), a French scientific expedition landed on Rodrigues, then occupied by the French. The scientific mission was headed by a well-known French astronomer and Catholic priest, the Abbé Alexandre Guy Pingré, who had been dispatched by the French Academy under the auspices of the expedition's patrons, Cardinal de Luynes and M. le Monier, to observe the transit of the planet Venus across the face of the Sun on June 6, 1761. The event was of extreme importance because, for the first time, it allowed for accurate measurements of both the orbit of Venus and the size of Venus as compared to the Sun. At this time, the best place for this study was on Rodrigues because of the favorable observation conditions: the Sun was visible at the time of the transit (the full transit takes between 20 and 40 minutes) and positioned high in the sky. In addition, Pingré and his fellow scientists had very precise instructions to gather specimens of such things as petrified shells, with the idea of comparing them with those found in Europe and other parts of the world.

Pingré had read Leguat's book and deliberately placed his observatory on the site of Leguat's settlement. Right away, he noted that the Huguenot traditions had not survived the trying times, since now "all those who live on Rodriguez make profession of being Christians; but each one after his own fashion," and thus were not religiously pure.[19] Indeed, Pingré disapprovingly observed, matters were such that slaves (on orders from the island's commandant) were attending prayers "offered by a slave who had never been baptized." Pingré also commented that "the work of Leguat passes for a fabric of fables, but here I found much less fables than what I was expecting."[20] Therefore, he used Leguat's journal as a roadmap to thoroughly survey the island. However, he never completed this survey, because the British seized Rodrigues and expelled the French.

Pingré nevertheless had time to look for the solitaire, if it still existed. A friend had told him that the birds were then not totally extinct, but they had become extremely rare and were only found in the most inaccessible parts of the island. Pingré did his best to scout these out, but he came back empty-handed. He never got to see a solitaire; chances are there weren't any left for him to see.

---

1 Another Marquis du Quesnes would, in the eighteenth
   century, become a leading commander in French
   Canada, giving his name to Fort Duquesne, which was
   renamed Pittsburgh after the British victory in the
   French and Indian War (1754–60). Interestingly and ironically, the name Duquesne is preserved in Pittsburgh as

the name of a Roman Catholic university founded in
1878.

2 Henri du Quesne, *Recueil de quelques memoires servans
d'instruction pour l'etablissement de l'isle d'Eden*,
Amsterdam: H. Desbordes, 1689.

3 François Leguat and Oliver Patsfield (ed.), *The Voyage of
François Leguat*, London: The Hakluyt Society, 1891.

4 *Ibid.*

5 *Ibid.*

6 *Ibid.*

7 *Ibid.*

8 *Ibid.*

9 *Ibid.*

10 *Ibid.*

11 *Ibid.*

12 *Ibid.*

13 François Cauche, *Relation du voyage que François
Cauche a fait à Madagascar, isles adjacentes & coste
d'Afrique, recueilly par le Sieur Morisot, avec das notes en
marge*, Paris: Roche Beullet, undated.

14 François Leguat and Oliver Patsfield (ed.), *The Voyage of
François Leguat*, London: The Hakluyt Society, 1891.

15 *Ibid.*

16 *Ibid.*

17 *Ibid.*

18 *Ibid.*

19 M. L. A. Milet-Mureau, *Voyage de La Pérouse autour du
monde, publié conformément au decret du 22 Avril 1791*,
Paris: Imprimerie de la République, 1797. See also
Alexandre Guy Pingré and J. Alby & M. Serviable (ed.),
*Courser Vénus: Voyage scientifique à l'île Rodrigues 1761,
fragments du journal de voyage de l'abbé Pingré*, Saint-
Denis de La Réunion: ARS Terres Créoles, 1993.

20 *Ibid.*

Chapter 6

The Rise of Dodology

THE DODO WAS BROUGHT TO EXTINCTION in more ways than one. First it was killed off in the flesh, then almost killed off in memory. If live specimens of the Mauritius dodo had not been taken back to Europe, the bird would not even have survived as an image. As it turned out, a string of unlikely circumstances kept the Mauritius dodo from becoming culturally extinct and nudged it toward its ultimate status as a unique and indispensable modern icon.

A major step along the way was the role it played as a scientific football in a momentous though little-known early battle over the theory of evolution. While the debate would have occurred even in the absence of this curious creature, both sides puzzled over and argued about the dodo.

But long before this, the dodo's bizarre anatomy, which was the prime factor in its physical demise, weighed just as heavily against it in the cultural realm. What was known about the dodo was unbelievable. There seemed to be no reliable records, only naïve drawings and vague accounts, remnants of a more credulous age when fantastic beasts adorned the maps as well as the memoirs of voyagers. The dodo was too much of a caricature to be real, and was relegated to the foggy realm of sailors' tales and maritime legend. By the early nineteenth century, most people who had heard of the dodo had no idea it ever really existed.

If anything, we owe the true story of these birds and how they lived to the efforts of a succession of often-contending naturalists, whom we can designate as the first serious dodologists. Spanning more than two centuries, their efforts to deal with the mysteries of the bird's genesis, adaptation, and disappearance would help set the stage for Charles Darwin's monumental *On the Origin of Species* in 1859. The sheer extravagance of the dodos of the Mascarenes would be crucial to the forging of a scientific view of evolution that different factions ultimately could endorse and thereby raise to near-universal consensus.

What bound these researchers together, yet fed their quarrels with each other as much as, if not more than, with outsiders, was their firm belief that the dodo had indeed existed. This belief was crucial to otherwise conflicting theories of adaptation and extinction. The clash would have its culmination in a passionate academic dispute that took place at Oxford University in the middle of the nineteenth century—a dispute that very likely inspired Charles Dodgson, aka Lewis Carroll, to include a dodo in Wonderland.

The first true naturalist to give credence to the dodo, making him perhaps history's first true dodologist, flourished long before the physical extinction of the dodo. This was the celebrated French-born scholar Carolus Clusius, who in his 1605 treatise *Exoticorum decem libris* published what is arguably the first scientific description of the Mauritius dodo. It was based on observations of remains of the bird, such as a foot preserved at the house of a friend, the anatomist Peter Paauw, combined with the study of ships' logs, wooden carvings, and tales from sailors. A pioneer of modern botany, Clusius (1516–1609) was the director of the Holy Roman Emperor's gardens in Vienna from 1573 to 1587, under the reign of Emperor Rudolf's father Maximilian II and several years into Rudolf's reign.

Clusius was especially interested in how plants from other parts of the world adapted to European conditions. While in

Vienna, he was given a collection of tulip bulbs by the Hapsburg ambassador to the Ottoman court in Turkey, where tulips had been cultivated for centuries. Clusius spent the later years of his life teaching at the university in Leiden, where he successfully cultivated tulips in 1593, thus beginning the Dutch bulb industry. His prose is much more polished than a sailor's account, and in his description, details of the dodo's morphology emerge in an organized fashion:

> The beak was thick and long, yellowish next to the head, with a black point. The upper mandible was hooked, the lower had a bluish spot in the middle between the yellow and black part, the bird was covered with thin and short feathers, the hinder part was very fat and fleshy, the legs were thick, covered to the knee with black feathers, the feet yellowish, the toes three before and one behind. Stones were found in the gizzards of these birds, and I saw two in Holland, one of which was about an inch in length.[1]

Significantly, Clusius was one of the first to give the Mauritius dodo a Latin name: *Gallus gallinaceus peregrinus*, which can be crudely translated as "foreign cock of the chicken family"—*peregrinus* originally meaning alien or foreign, but later meaning wandering (it is the root of the word "pilgrim"). When a stuffed specimen arrived at Oxford University's Ashmolean Museum in 1683, it was cataloged as "No. 29 *Gallus gallinaceus peregrinus*, Clusii," i. e., the name given it by Clusius nearly a century earlier.[2] Meanwhile, in *A Catalogue and Description of the Natural and Artificial Rarities Belonging to the Royal Society*, published in 1681, the respected naturalist and physician Nehemiah Grew wrote the following under the drawing of a leg: "The leg of a Dodo; called *Cygnus cucullatus* [cuckoolike swan] by Nierembergius; by Clusius, *Gallus gallinaceus peregrinus*; by Bontius called Dronte, who saith that by some it is called (in Dutch) Dodo-aers."[3] It appears that the term *dronte* is originally Dutch, though it would become the standard French word for dodo. It is not clear who Nierembergius was, though there is

This is the dodo presented in Clusius's Exotica *(1605), and there-fore generally referred to as "the dodo of Clusius." It is a female in the height of moult. One of two gizzard stones examined by Clusius at the home of a friend in Leiden is also shown and described as being an inch in length.*[4] (From Hugh Edwin Strickland, *The Dodo and Its Kindred; or, The History, Affinities, and Osteology of the Dodo, Solitaire, and Other Extinct Birds of the Islands Mauritius, Rodriguez, and Bourbon,* London: Reeve, Benham, and Reeve, 1848. From the collections of the Ernst Mayr Library of the Museum of Comparative Zoology, Harvard University.)

A FIERY STOMACH:

It is a slow-paced and a stupid bird, and which easily becomes prey to the fowlers. The flesh, especially the brest, is fat, esculent, and so copious, that three or four Dodos will sometimes suffice to fill a hundred seamen's bellies. If they be old or not well boiled they are of difficult concoction, and are salted and stored up for provi-sion of victual. There ar found in their stomachs of an ash color, of diverse figures and magnitudes; yet not bread there, as the common people and seamen fancy, but swallowed by the bird; as though by this mark also would manifest that these fowl are of the ostrich kind, in that they swallow any hard things but do not digest them.[5]

mention of a Spanish Jesuit Latinist named Johann Eusebius Nierembergius. Bontius is possibly Jacob Bontius, a seventeenth-century professor of medicine at Leiden and an expert on the flora and fauna of the East Indies. (He is said to have made the first descriptions of such diseases as beriberi and rickets.)

The same leg is mentioned again in the *Onomasticon* (a kind of dictionary of names) by the English physician and naturalist Walter Charleton (1620–1707), published in 1688. Here, the author refers to the bird as "*Dodo lusitanorum*, the *Cygnus cucullatus* of Willughby and Ray."[6] Francis Willughby (1635–1672) and his teacher at Cambridge University, John Ray (1627–1705) were a celebrated team of naturalists noted for their books on birds.

Gallus peregrinus? *Cygnus cucullatus*? *Lusitanorum* (of the Lusitanians, i. e., the Portuguese)? By the eighteenth century, there was a proliferation of "scientific" names applied to the Mauritius dodo. Where did they come from? Why are they so different from one another?

―――――――――

It was timing that made the dodo a pivotal figure in the shaping of our modern culture, coming as it did just as taxonomy emerged as an important scientific discipline that classifies life in terms of form and behavior. During the course of the eighteenth century, taxonomy became solidly established in the study of the natural sciences as a means of dealing with the rapidly growing diversity and number of life-forms from all over the globe being brought to the attention of naturalists. The discipline is concerned with organizing the entire living world inside precise brackets of closer or looser proximity, depending on the morphology, physiology, and all-around behavior of each species. However, by the time taxonomy started placing living things into distinct groups of related individuals, the dodo of Mauritius, the white dodo of Réunion, and the solitaire of Rodrigues were extinct, even though their various Latin names were well-established. While some naturalists claimed these bizarre creatures simply never existed, others went to work trying to find them a taxonomical niche.

Classifying the dodo in scientific terms was a task as hard as understanding the dodo's lifestyle and the factors accounting for its anatomy. Dodologists had to think hard and debate even harder to persuade fellow scientists of their points—and, in the process, as it happened in so many other cases, science evolved faster and with better results than it would have otherwise.

Years after Clusius called the dodo *Gallus gallinaceus peregrinus*, the dodo of Mauritius was christened *Raphus cucullatus*. *Raphus* was the invention of the German naturalist Paul Heinrich Gerhard Moehring, meant as a Latinization of the Dutch *reet*, a vulgar term for rump. *Cucullatus* was simply a more dignified way of saying cuckoo. Thus, this designation stands loosely for something akin to a "cuckoo-like bird with fat rump."

Now, the actual dodo was nothing like a cuckoo, though apparently Moehring thought otherwise. Moreover, it didn't have a large rump. Therefore, the preeminent Swedish taxonomist and botanist Carolus Linnaeus (1707–1778), who had established the binomial, or two-name, system of classifying flora and fauna, decided to give the poor creature a proper name. He came up with *Didus ineptus*, something to the effect of "clumsy dodo." This new designation, being Linnaeus's own, carried so much weight with his contemporaries that most naturalists promptly adopted it, dropping the earlier terms. But not everyone played along. After several decades of the use of *Didus ineptus*, the holdouts who favored using *Raphus cucullatus* joined their voices in protest, for one of the golden rules of taxonomy was that the first Latin name given to a species is the one that prevails, a rule Linnaeus himself had enforced often enough during his long years of fighting for order in the taxonomical universe. And so the clumsy dodo became a cuckoo with a large rump again. Since the white dodo of Réunion was obviously a close relative of the darker dodo of Mauritius, it was a good candidate for Linnaeus's *Didus ineptus* tag. However, since the white bird was obviously a different species of the same genus, the name was changed to *Didus borbonicus*, reflecting the fact that, at that time, Réunion went by the name of Bourbon. Therefore, the second name appeared to function as an homage to the Bourbons, France's royal family. Bearing in mind, however, the characteristically dumb face of the fat, slow dodo, one wonders how innocent an expression of reverence this change was intended to convey, given the decline of the house of Bourbon after the passing of Louis XIV.

As for the Rodrigues solitaire, no one knew exactly what to make of it. It ended up being classified as a different genus of the same family and referred to by the vague name of *Pezophaps solitarius*, the "solitary walking pigeon." The family including the three ill-fated birds was called *Raphidae*, after the Mauritius dodo. Linnaeus was simultaneously a culprit and a victim of the confusion surrounding the first attempts at properly classifying the Mauritius dodo and its cousins.

While chronicling his exiled life at Rodrigues, François Leguat had written about the solitaire. He also described what he referred to as a "giant": a creature six feet tall, with long legs and a tiny body. This could have been a crane or a heron, but one of Leguat's contemporaries called it a kind of ostrich. The erroneous idea thus spread that Leguat's solitaire was a form of ostrich with very short legs. And so, when imposing order in the animal kingdom during the eighteenth century, and certainly with the best of intentions, Linnaeus put the dodo of Mauritius and its kin down as short-legged ostriches.

The last known dodo—the stuffed specimen from the collection of the naturalist John Tradescant donated to Oxford's Ashmolean museum in 1683—was at least a century old when it was tossed out on January 8, 1755. Fortunately, someone removed the head and a foot of the specimen and saved them. The rest was burned as trash.

As noted in the museum's records, the removal of the dodo and several other specimens was ordered "at a meeting of the majority of the visitors."[7] The order was decreed by the vice-chancellor George Huddesford and his trustees, passing around a list at their annual meeting, as stipulated by Elias Ashmole, the museum's original benefactor. If they knew they were destroying the last dodo in existence, dead or alive, they did not care. As the great geologist Charles Lyell wrote almost a century after the event,

> Some complain that inscriptions on tombstones convey no general information except that individuals were born and died—accidents which happen alike to all men. But the death of a *species* is so remarkable an event in natural history, that it deserves commemoration; and it is with no small interest that we learn from the archives of the University of Oxford the exact day and year when the remains of the last specimen of the Dodo, which had been permitted to rot in the Ashmolean Museum, were cast away.[8]

At the same time, an influential coterie of naturalists was giving the dodo a new lease on intellectual life. Chief among these was Georges-Louis Leclerc, Comte de Buffon (1707–1788), a wealthy French aristocrat who was the most prolific and best-known science popularizer of his time. A mathematical genius, at the age of 20 he is said to have discovered the binomial theorem. Among his many accomplishments were turning Paris's royal botanical garden, the *Jardin des Plantes*, which he managed from 1738 until his death, into a genuine scientific enterprise, and the publication of the 44-volume *Natural History*, the most widely read scientific compendium of the day. In the section devoted to birds, published between

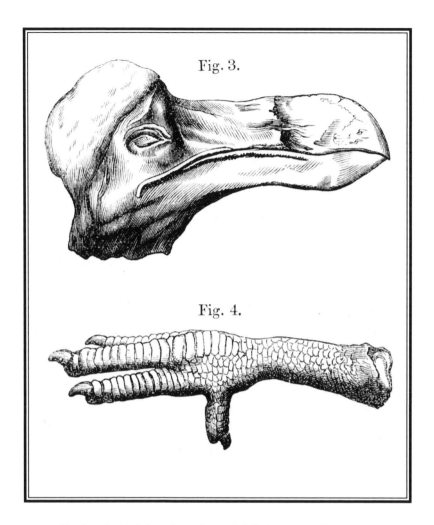

Fig. 3.

Fig. 4.

Having decided that the only stuffed dodo in its collection was no longer presentable, the Ashmolean Museum at Oxford burned it with other trash. Only this head and foot were rescued for posterity. It was the last dodo specimen left in the world. (From Richard Owen, *Memoir of the Dodo* (Didus ineptus, Linn.), London: Taylor and Francis, 1866. From the collections of the Ernst Mayr Library of the Museum of Comparative Zoology, Harvard University.)

1770 and 1783, Buffon helped immortalize the dodo, known in French as *dronte*.

> In most animals, weight means strength; but here it produces
> only heaviness; the ostrich or the casoar are not more able to
> fly than the dronte, but at least they are very fast at the race;
> but the dronte seems immobilized by its own weight, & to be
> able only of dragging itself; it seems to be composed of a
> brute, inactive matter, where not enough living molecules
> were used; it has wings, but these wings are too short and too
> weak to take it up in the air; it has a tail, but this tail is
> disproportionate and out of place; it looks like a turtle that
> wrapped herself in the remains of a bird, & Nature, by giving
> it these useless frills, seems to have wanted to top them with
> the embarrassment of excessive weight, the ineptitude of the
> movements, & the inertia of the mass, & to make its heavy
> thickness even more shocking because it belongs to a bird.[9]

Not only did Buffon push back the age of the earth to some 75,000 years, he also proposed an eighteenth-century model for evolution. It was a very smart and solid model, and it would survive Darwin's ideas for at least three decades after the publication of *On the Origin of Species*, fueling one side of a celebrated debate over Darwinism that took place at Oxford in 1860. In effect, Buffon reasoned, it was animal degeneration that led to the formation of new species.

In this view, animals are shaped over time by the sculpting powers of their environment. Buffon's prime example was man himself, once a homogeneous prototype but now broken down into a multitude of different, "degenerate" races. As he put it in his chapter "Degeneration of Animals," included in the "Quadrupeds" section of his *Natural History*:

> The changes became so big and remarkable that we would
> have believed that the black, the Laplander, and the white,
> form three different species, if it weren't for the fact that we
> are sure that only one man was created, and also because the
> black, the Laplander, and the white, can become united and
> propagate together the big and unique family of our human
> kind: the blood is different, but the germ is the same.[10]

In his capacity for breeding and interracial propagation, man demonstrates an elasticity way beyond what is found in animals, and even more so than in plants; according to Buffon, this elasticity comes "less from man's body than from his very soul."

This elasticity allows us to gain further insight into the history of the earth: In Buffon's scheme of things, the earth was divided into two continents, the "old" and the "new," i. e., Eurasia-Africa and the Americas. It was a division "older than all our monuments; but man is even older, because he can be found in the two worlds."

Through deliberate breeding, changes could be dramatically accelerated, while the effect of the environment is much slower. Buffon thought it would take

> but one hundred and fifty or two hundred years to wash the skin of a black through the mixture with the blood of a white; but we would need most likely just as many centuries to achieve the same effect through the influence of climate alone. . . .

But precise experiments in this realm would be hard to bring off:

> We would have to transport some individuals from this black race from Senegal to Denmark. We would have to enclose these blacks with their females, and to conserve the race carefully, without allowing any mixtures; this means is the only one we can employ to define how much time we would need to reintegrate the nature of white man and how long it took to change from white to black.[11]

In addition, sexual differences had to be kept in mind when it came to breeding to perfect a race: "The strength of the body and the bigness of size are male characteristics; demeanor and beauty are female characteristics." Other influences could also produce different physical manifestations. For instance, Buffon insisted, people ought not neglect the importance of feeding habits, since food is the way through which the land shapes the body, just as the sky shapes the species by changing the skin. The nutritional factor was more important for herbivores than carnivores, since plants directly transmitted the properties of the land:

> In general, the influence of the food is bigger, and has stronger effects, on the animals that live on herbs and fruit; those that, on the contrary, live only on their prey vary less due to this cause, and more due to the climate's influence; because meat is a food already prepared and already assimilated to the nature of the carnivorous animal that devours it; whereas the herb is a primary product from the land, it has all of the land's properties, and immediately transmits terrestrial properties to the animal feeding on it.[12]

Buffon illustrated this concept by comparing European sheep breeds to those in rest of the world: "They all present the mark of their environment: in some aspects they became more perfect, in others they show more vices. But since to perfect or to show vices are the same thing for nature, they are all degenerated, because they all were changed."

Buffon noted that the factors of land and sky have a more immediate effect on animals than on man because animals, unlike man, tend to have similar eating habits, are more closely dependent on the land, and cannot build huts or don coats, or even light a fire, and therefore are much more exposed to the elements. This is why, among animals, the representatives of a species tend to appear in one place: they are helpless against the environment, and it is the place that suits them best.

Together with "the temperature of the climate" and "the quality of the food," Buffon pointed out a third cause responsible for the degeneration of animals: "the evils of slavery." According to this thesis, enslaved animals end up even losing some of their original natural traits, becoming unable to survive on their own. Buffon illustrated his idea that slavery can lead to transmissible deformities with the example of the lowly camel:

It is born with humps on its back, and with callosities on its legs and on its chest: his callosities are evident wounds caused by constant rubbing, since they are full of pus and corrupted blood. Since it always walks with a heavy load, the pressure from the harness has first made [it] impossible for the muscles in the back to grow uniformly, and then it made the flesh of the neighboring areas become swollen; and since, whenever the camel wants to rest and sleep, they make it first fall on its hindlegs, and since the animal has slowly adopted this habit itself, all the weight of its body rests several hours in a row, every day, over its chest and its knees; and the skin of these parts is pressed, rubbed against the sand, hardens, and becomes disorganized.[13]

As for "free" animals, less dependent on food and climate and spared man's intervention, variation comes mainly from their mating habits: those species that are monogamous and reproduce slowly have fewer variations; those that change partners often, have several pups per litter, and breed more than once a year vary the most. This seems to occur in direct correlation with the size and the perfection of the species: big, complex animals breed slowly, whereas small, inferior animals breed quickly.

Strong similarities between animals in the Old and New Worlds led Buffon to surmise that they were not the product of the process of degeneration pure and simple. In his opinion, the similarities proved that the two continents were once united, becoming separated by the Atlantic only after animals sprang from the same original source. Their shaping by the elements then followed different courses, creating the present-day species only after the waters had separated the land. Buffon added that after this separation, "American creatures became smaller and more degenerated," an argument that would enrage Thomas Jefferson.[14]

This whole process of speciation through degeneration might seem somewhat prodigious now, in the sense of being outlandish. However, Buffon was quick to remind his readers: "What is a prodigy, in Nature, other than an effect rarer than the others?"[15]

The growth of taxonomy under Linnaeus and the influence of thinkers like Buffon—and his protégé and successor as director of the *Jardin des Plantes*, Jean-Baptiste Lamarck (1744–1829), who believed that there was no such thing as extinction, only evolution into other forms—gave a boost to maritime exploration for its own sake, with naturalists as crew members and the mission of collecting specimens of flora and fauna, and fossils. Captain James Cook's voyages in the Pacific were an example.

But the most famous of all would be the exploration of the coasts of South America by the British naval ship *Beagle* with the young naturalist Charles Darwin on board. Interestingly enough, in April 1836 the *Beagle*, on the final leg of its five-year voyage around the world, made a stop at Mauritius, which Darwin described as "adorned with an air of perfect elegance."[16] The Mascarene Islands had been taken over by the British in 1811 during the war against Napoleon, and Mauritius was undergoing a boom fueled by the cultivation of sugar cane. Darwin, starved of the creature comforts of civilization, was impressed by the French-style capital, which had an opera house and tarred roads, and he toured the island riding on an elephant provided by his English host. One can guess that he heard of the dodo, but there is no mention of it in his account of Mauritius, nor, for that matter, anywhere in his writings.

Nevertheless, the main contention during the debate that would take place at Oxford over the dodo was centered on the ideas Darwin himself was quietly marshaling for his groundbreaking work *On the Origin of Species*. If you were a naturalist in that period, you had to take sides. Either you endorsed what would become known as the Darwinian concept of evolution in a cautious or an ardent manner (it didn't matter which), or you rejected it altogether and took a stake in the defense of previous explanations for the diversity of life, such as those proposed by Buffon or Lamarck. Or else you had to come up with your own views, and make them consistent and resilient in the face of adverse data.

The dodo was ideal for a controversy, because its bizarre physiognomy lent itself to sharply divergent arguments about the causes that allow new species to arise from pre-existing forms. Those who would later take Darwin's side saw in the dodo the confirmation that some species could arise out of juvenile forms that became sexually mature, and spend entire lifecycles without ever developing the adult characteristics, due to peculiarities of their environment. This phenomenon, well documented in numerous organisms, is known as *neoteny*. On the other hand, the dodo was an exemplary vehicle for the colorful ideas of species formation put forth with masterly eloquence by Buffon. As we have seen, Buffon preferred to think of speciation as a phenomenon occurring through different pathways of degeneration, imposed on the original stock by environmental conditions and breeding habits.

Facing an audience of skeptical peers inclined to dismiss the dodo as folklore or a flight of the imagination, the Oxford polemicists resurrected the bird from the only evidence available to them: dispersed bones, partial skeletons, the diaries of seventeenth-century sailors and accompanying woodcut illustrations, and the small collection of perplexing dodo portraits by artists like Hoefnagel and Savery.

The paintings, drawings, and engravings were dispersed among private owners, antiquarians, and random libraries. This made them hard to track down, pull together, and study as a whole. More often than not, the depictions of the dodo in works of art were not set in naturalistic backgrounds, but rather represented odd collections of rarities gathered in the age of discovery, then assembled in mythological or Eden-like compositions dictated by the tastes of wealthy patrons like Rudolf II. Thus, dodos were portrayed in the company of creatures from all over the world, doing things a real dodo would not. For instance, in one of Savery's paintings, a dodo is shown standing by a small creek and staring avidly at what seems to be a small eel in the water. This image triggered a long debate in the nineteenth-century about whether or not dodos could have been carnivores.

Therefore, in order to make sense of their observations, the dodologists of Oxford had to lift the dodos from the misleading contexts of the original paintings and reassemble them. In these reconstructions, the bird was sometimes shown alone. Sometimes details of the flora that should have been there were drawn around them, or else the animals and plants that accompanied them in the original work were simply outlined, so that the viewer would know which particular creature was under consideration. For all their good intentions, these procedures simply created additional confusion.

But how could a dozen bones and some brushstrokes result in reliable science? It is hard to imagine scientists taking up such a challenge today. The early nineteenth-century efforts to reconstruct the dodo are even more impressive if we bear in mind that the brief account of a gentleman named Sir Hamon Lestrange, who saw a live dodo displayed to a paying public in London, was for awhile the only eyewitness. At that time, no additional bones had been found in the Mascarenes. Several amateur naturalists had spent their lives on the islands and amassed large bone collections, but none of these included dodo remains.

It was erroneously assumed that, if the dodo had inhabited Mauritius, the best place to look for its remains would be nearby Madagascar, since this was a much larger island with abundant fauna. All searches carried out on Madagascar, however, proved to be fruitless. Earlier, lengthy descriptions of Madagascar's ecosystems by seventeenth-century travelers never mentioned a dodo-like creature. This persistent absence of evidence fed skepticism about a bird that no scholar had ever seen and no native folklore had ever regis-

Five of the dodo oil paintings where known to exist when Strickland wrote The Dodo and Its Kindred. *One was anonymous, three were by Roelandt Savery, and one by Roelandt's nephew Hans Savery. The most famous of Savery's paintings depicts Orpheus charming the animals with his music, and among innumerable birds and beasts, all depicted with the utmost accuracy, we see the clumsy dodo spellbound by the strains of the lyric bard. This painting contains one of the most animated dodos ever painted, which led Strickland to believe that it had been drawn from a live specimen. It is the famous bird looking at the eel in the water that led to the argument about whether dodos could have fed on small aquatic creatures.* (From Richard Owen, *Memoir of the Dodo* (Didus ineptus, Linn.), London: Taylor and Francis, 1866. From the collections of the Ernst Mayr Library of the Museum of Comparative Zoology, Harvard University.)

In order to better understand their subject matter, nineteenth-century dodologists often removed the Mauritius dodos from the original paintings for the sake of realism. These dodos have been cut and pasted from three different paintings. The dodo on the left is the one that delighted William John Broderip "in all of the beauty of its ugliness."[7] (From Richard Owen, *Memoir of the Dodo* (Didus ineptus, Linn.), London: Taylor and Francis, 1866. From the collections of the Ernst Mayr Library of the Museum of Comparative Zoology, Harvard University.)

tered. It had taken the dodo around a century to die after its discovery. Another century and a half would be required for the dodo to reemerge through words, pictures, and a handful of spare parts.

———————

Breakthrough efforts in young sciences involve ingenuity. Naïve as the approach might seem to us now, the painstaking analysis of old paintings bore fruit. Not only was the existence of the dodo validated through them, but the effort required the dodologists to pour over a diverse pool of evidence, including sailors' accounts, old wood carvings, and second-hand accounts that described the Mascarenes' flora and fauna, in order to develop a more complete understanding of the bird.

In *The Dodo and Its Kindred*, published in 1848, Hugh Edwin Strickland, the president of the Ashmolean Society in Oxford (repository of two pathetic dodo remnants), working in collaboration with his coadjutor Alexander Melville, began a formidable effort to bring the dodo back to life and to understand and explain its extinction.

In addition to the Ashmolean remnants, Strickland had examined dodo bones found in a cavern on Rodrigues by Mauritian naturalists and exhibited by Georges Cuvier in his museum in Paris. Cuvier (1769–1832) was a follower of Linnaeus and heir to the mantle of Buffon and Lamarck as France's leading man of science. On man's effect on Nature, Strickland wrote:

> It appears, indeed, highly probable that Death is a Law of Nature in the Species as well as in the Individual; but this internal tendency to extinction is in both cases liable to be anticipated by violent or accidental causes. Numerous external agents have affected the distribution of organic life at various periods, and one of these has operated exclusively during the existing epoch, viz. the agency of Man, an influence peculiar in its effects, and which is made known to us by testimony as well as by indifference. The object of the

present treatise is to exhibit some remarkable examples of the extinction of several ornithic species, constituting an entire sub-family, through Human agency, and under circumstances of peculiar interest.[18]

Practically nothing is now known of Strickland, not even the year of his birth, but in his book on the dodo we see him, a decade before the publication of Darwin's *On the Origin of Species*, reasoning on distribution patterns and species adaptation in a tentative portrait of the ecosystem:

> We find a special relation to exist between the structures of organized bodies and the districts of the earth's surface which they inhabit. Certain groups of animals and vegetables, often very extensive, and containing a multitude of genera and species, are found to be confined to certain continents and circumjacent islands.[19]

This he illustrated in a concise footnote: One example "among a thousand," the hummingbird group contains hundreds of species and is exclusively confined to the Americas and the West Indies. Having made this point, the author continued:

> In the present state of science we must be content to admit the existence of this law, without being able to enunciate its preamble. It does *not* imply that organic distribution depends on soil and climate; for we often find a perfect identity of these conditions in opposite hemispheres and in remote continents, whose faunae and florae are almost wholly diverse. It does *not* imply that allied but distinct organisms have been educed by generation or spontaneous development from the same original stock; for (to pass over other objections) we find detached volcanic islets which have been ejected from beneath the Ocean (such as the Galapagos for instance) inhabited by terrestrial forms allied to those of the nearest continent, though hundreds of miles distant, and evidently never connected with them. But this fact *may* indicate that the Creator in forming new organisms to discharge the functions required from time to time by the ever vacillating balance of Nature, has thought fit to preserve the regularity of the System by modifying the types of

structure already established in the adjacent localities, rather
then to proceed per saltum [by leaps] by introducing forms of
more foreign aspect.[20]

After setting the stage for the extinction of the "ornithic
species" that he held so dear, Strickland began describing the
dodos: "These birds were of large size and grotesque proportions, the wings too short and feeble for flight, the plumage
loose and decomposed, and the general aspect suggestive of
gigantic immaturity." Their history was every bit as unique.
Two centuries before, men had colonized their native islands
and "speedily exterminated" them. "So rapid and so complete was their extinction that the vague descriptions given of
them by early navigators were long regarded as fabulous or
exaggerated, and these birds, almost contemporaries of our
great-grandfathers, became associated in the minds of many
persons with the Griffin and the Phoenix of mythological
antiquity." Strickland declared it was his aim "to vindicate the
honesty of the rude voyagers of the seventeenth century."

Calling these birds *Didinae*, Strickland plunked down on
the side of Linnaeus in the taxonomical debate. He stressed
that he had not picked this group by chance or simply
because of its bizarre aspect; rather, he reminded us, these
doomed creatures stood in human history as a powerful
symbol. The *Didinae* furnish the first clearly attested
instances of the extinction of organic species by human
agency. Pondering at length the meaning of such vanishings,
Strickland invoked the special relation of God and man:

> Our consolation must be found in the reflection, that Man is
> destined by his Creator to "be fruitful and multiply and replenish
> the Earth and subdue it." The progress of Man in civilization, no
> less than his numerical increase, continually extends the
> geographical domain of Art by trenching on the territories of
> Nature, and hence the Zoologist or Botanist of future ages will
> have a much narrower field for his researches than that which we
> enjoy at present. It is, therefore, the duty of the naturalist to
> preserve to the stores of Science the knowledge of those extinct or
> expiring organisms, when he is unable to preserve their lives.[21]

"The oldest inhabitants assure everyone that these monstrous birds have been always unknown to them. However this may be, it is certain that for nearly a century no one has here seen an animal of this species. But it is very probable that before the islands where inhabited, people might have been able to find some species of very large birds, heavy and incapable of flight, and that the first mariners who sojourned there soon destroyed them from the facility from which they were caught."[22] (From Richard Owen, *Memoir of the Dodo* (Didus ineptus, Linn.), London: Taylor and Francis, 1866. From the collections of the Ernst Mayr Library of the Museum of Comparative Zoology, Harvard University.)

It would be no simple task, or as Strickland put it: "The paleontologist has, in many cases, far better data for determining the zoological characters of a species which perished myriads of years ago, than those presented by a group of birds, several species of which were living in the reign of Charles the First." Strickland then reiterated his belief that the dodo was real:

> Most persons are acquainted with the general facts connected with that extraordinary production of Nature, known by the name of the Dodo—that strange abnormal Bird, whose grotesque appearance, and the failure of every effort made for the last century and a half to discover living specimens, long caused its very existence to be doubted by scientific naturalists. We possess, however, unquestionable evidence that such a bird formerly existed in the small Island of Mauritius, and it is ascertained with no less certainty that the species has been utterly exterminated for a period of nearly two centuries.[23]

Assuming that the dodo was indeed real, then where did its strange morphology come from? This fundamental question would be at the core of a dispute between Strickland and two fellow ornithologists, Richard Owen and William John Broderip, who joined efforts in a very impressive compilation cleverly titled *Memoir of the Dodo* in 1866. Broderip collected transcripts from old accounts and reunited them in a long preface, while Owen took upon himself the painstaking task of measuring, comparing, and drawing for our instruction all the available bones of dodos in museums.

Whereas outside dodology Strickland became virtually unknown to posterity, Owen is a very different case. Born in Lancaster, England, in 1804, Owen was trained as a surgeon and began his career as a naturalist in London when he became general assistant to William Clifts, conservator of the Hunterian Collection in the Royal College of Surgeons. In 1830, Clifts asked him to take the great French naturalist

Georges Cuvier on a tour of the Hunterian Collection when Cuvier visited England. Cuvier reciprocated by inviting Owen to visit him in Paris the following year; there, Owen carefully studied Cuvier's specimens in the National Museum of Natural History. Until his death in 1892, Owen would refer to this study as "the major influence on my work."[24]

Upon Clift's death, Owen took his place and, in 1856, was appointed superintendent of the natural history departments in the British Museum, for which he designed new quarters in South Kensington. At this point, he left medical practice and devoted himself to research. He had already begun dissecting dead animals from the gardens of the Zoological Society in London in 1828, and during the rest of his career, he amassed major awards in comparative anatomy, vertebrate paleontology, and geology. He helped delineate several natural groups and first described many organisms, both recent and fossil. Indeed, he is credited with coining the term "dinosaur." He published dozens of papers, which earned him the reputation of being "the English Cuvier." His work in comparative anatomy and paleontology was truly in the best Cuverian tradition. Among his many honors was being given a house in Richmond Park by Queen Victoria herself, and upon his retirement he was knighted with the Order of the Bath.

In the course of his studies, Owen became fascinated by monotremes (egg-laying mammals like the platypus) and marsupials (mammals like kangaroos that carry their newborn in pouches), either live or in fossil form. Numerous expeditions sent to Australia and New Zealand returned specimens to England for examination by Owen. Likewise, Owen was passionately interested in primates, primarily anthropoid apes, and their relation to human beings. Again, several African explorers sent him specimens for study and classification.

For Owen, the culmination of these explorations came in 1839, when an expedition working in New Zealand under his direction discovered a femur that appeared to belong to a previously unknown giant bird, the creature now known as the

New Zealand moa. This finding would be of extreme relevance to the taxonomical debate over the dodo, in which Owen was, characteristically, ever eager to carve out, defend, and vindicate his own opinions.

———————

Today, Owen is remembered as much for his strong opposition to the views of Charles Darwin as for his many contributions to the study of fossil animals. Owen claimed that he opposed Darwin not on the question of evolution but on the mechanism of natural selection, and he put forth complicated views on "transmutations of species" that are not exactly clear, partly because of his writing style. In 1848, he declared he could think of six ways in which the Creator might have acted to create species, but he would not enumerate them. This led to a debate with one of Darwin's chief defenders, Thomas H. Huxley, many years Owen's junior, which culminated in 1860 with a somewhat pathetic (for Owen) public confrontation between the two men during a meeting of the British Association for the Advancement of Science—of which Owen had been president in 1858, but whose members now seemed to turn against him. His bitterness over this public embarrassment seemed to lead Owen to coach the Bishop of Oxford, Samuel Wilberforce, in the latter's endless polemics against Huxley. Ironically, Owen and Darwin had been colleagues for 20 years before their differences exploded. Owen was introduced to Darwin by Charles Lyell, and the two men started a friendly collaboration when Darwin asked Owen to help him classify his South American fossils.

We can only assume that either Owen's judgment was muddied by his sense that his own preeminence in biology was about to be lost when *On the Origin of Species* was published in 1859, or he may have been so strongly attached to the ideas inherited from Buffon that any alternative explanation of evolution was heresy. Like Buffon, Owen believed that the

*With only small fragments preserved and only a few paintings
and engravings to speak for the crazy bird, nineteenth-century
dodologists had to piece together the creature of their study
from fragmentary evidence. The skeleton in this plate is recon-
stituted from separate bones, and used as a scaffold for the surrounding silhouette.* (From Masauji Hachisuka, *The Dodo
and Kindred Birds; or, The Extinct Birds of the Mascarene
Islands*, London: H. F. & G. Witherby, Ltd., 1953.)

dodo had been a misguided and hasty creation of a juvenile and unwise habitat. Like Buffon, Owen assumed that Mauritius was geologically too young and too small to come up with less bizarre and more intelligent animal productions. To both men, the disaster called the dodo displayed ineptitude on all planes, even its mating habits. According to Owen, the dodo's monogamous character deprived the bird of "the excitement, even, of a seasonal prenuptial combat," allowing it to "go on feeding and breeding in a lazy, stupid fashion," thus blocking "any growth of the cerebrum proportionate to the gradually accruing increment of the bulk of the body."[25]

Now intent on discrediting Darwin, Owen wrote a long anonymous review of *On the Origin of Species* published in *The Edinburgh Review* in 1860, which Darwin called "extremely malignant, clever, and I fear . . . very damaging. He misquotes some passages, altering words within inverted commas. It requires much study to appreciate all the bitter spice of the many of the remarks against me."[26]

As Darwin's theory became more accepted in the scientific community, Owen shifted his position somewhat. Although he denied Darwin's doctrine to the end, he admitted the accuracy of its basis, claiming to have been the first to point out the truth of the principle on which it was founded. Needless to say, the two men were never close again.

Strickland, on the other hand, had sided with Darwin from the beginning, holding that the dodo looks like a duckling suddenly blown up to gigantic proportions, which he interpreted thus: "The dodo affords one of those cases, of which we have so many examples in Zoology, where a species, or part of the organs in a species, remains permanently in an undeveloped or infantine stage."[27]

To reinforce his point, Strickland brought up other examples of neoteny, including the Greenland whale, "a permanent suckling," and the proteus, a microorganism that Strickland called "a permanent tadpole." Likewise, the dodo would have been "a permanent nestling . . . clothed with down instead of feathers, and with the wings and tail so short and feeble as to be utterly unsubservient to flight."[28]

But Owen, in the conclusion of *Memoir of the Dodo*, referred to those who, like Strickland, mentioned incipient organs required by evolution as "stigmatizers of Buffon."[29] Owen mocked Strickland's passages quoted above, then asked what the stigmatizers of Buffon had to offer in lieu of his theory as to the origin of the dodo. In response, Strickland asserted that:

> It may appear at first sight difficult to account for the presence of organs which are practically useless. Why, it may be asked, does the Whale possess the germs of teeth which are never used for mastication? Why has the Proteus eyes when he is especially created to dwell in darkness? And why was the Dodo endowed with wings at all when those wings were useless for locomotion? These apparently anomalous facts are really the indications of laws which the Creator has been pleased to follow in the construction of organized beings; they are inscriptions in an unknown hieroglyphic, which we are quite sure mean *something*, but of which we have scarcely begun to master the alphabet. There appear, however, reasonable grounds for believing that the Creator has assigned to each class of animals a definite type of structure from which He has never departed, even in the most extravagant and eccentric modifications of form. Thus, if we suppose that the abstract idea of a Mammal implied the presence of teeth, the idea of a Vertebrate the presence of eyes, and the idea of a Bird the presence of wings, we may then comprehend why in the Whale, the Proteus, and the Dodo, these organs were merely suppressed, not wholly annihilated.[30]

For Owen, this was laughable: "This notion of type-forms or centers, unfortunately, has not merely relation to abstract biological speculations or theories, but to practical questions on which the true progress of Natural History vitally depends. If such types do exist, the National Museum, they argued, may be restricted to their exhibition."[31]

To strengthen his argument against Darwin, Owen quoted from his own letter to *The Times* of London, written in May 1866.

> Some naturalists urge that it is only necessary to exhibit the type-form of each genus or family. But they do not tell us what is such "type-form." It is a metaphysical term, which implies that the Creative Force had a guiding pattern for the construction of all the varying or divergent forms in each genus or family. The idea is devoid of proof; and those who are loudest in advocating the restriction of exhibited specimens to 'types' have contributed least to lighten the difficulties of the practical curator in making the selection. . . .[32]

The dodo exemplifies Buffon's idea of the origin of species through departure from a more perfect original type by degeneration. In accord with Lamarck's idea, the known consequences of the disuse of one locomotive organ and extra use of another point to the secondary causes that may have operated in the creation of this bird. The young of all doves are hatched with wings as small as in the dodo, but the latter species retains the immature character. The main condition making possible the production and continuance of such a species on the island of Mauritius was the absence of any animal that could kill a great bird incapable of flight. The introduction of such a predator became fatal to the species, which had lost the means of escape. The Mauritian doves, which retained their powers of flight, continue to exist there.

But Strickland would have nothing of the ideas inherited from Buffon, especially the notion that imperfection or degeneracy could be involved:

> Each animal and plant has received its peculiar organization with the purpose, not of exciting the admiration of other beings, but of sustaining its own existence. Its perfection, therefore, consists, not in the number or complications of its organs, but in the adaptation of the whole structure to the external circumstances in which it is destined to live. So every

department of organic creation is equally perfect; the humblest animalcule, or the simplest conferva, being as completely organized with reference to its appropriate habitat, and its destined functions, as Man himself, who claims to be lord of all. Such a view of creation is certainly more philosophical than the crude and profane ideas entertained by Buffon and his disciples, one of whom called the Dodo *un oiseau bizarre, dont toutes les parties portaient le caractère d'une conception manquée* ["a bizarre bird, whose parts bore the characteristics of a failed conception"]. He fancies that this imperfection was the result of the youthful impatience of the newly formed volcanic islands which gave birth to the Dodo, and implies that a steady old continent would have produced a much better article.33

And this is Owen's final retort: "Nevertheless, the truth must be said. The *Didus ineptus*, through its degenerate or imperfect structure, however acquired, has perished."34 In other words, it failed to adapt.

———

Though both sides in the great debate over evolution agreed that there had been such a thing as the dodo and it was extinct, there remained the task of classifying the bird, resolving the absurdities engendered by eighteenth-century taxonomy. By the mid-1800s, opinions were still sharply divided— to the extent that Richard Owen differed with his beloved mentor Cuvier. They remained cordial friends until Cuvier's death, two years after Owen's trip to Paris. However, Cuvier stubbornly held to the idea that the dodo belonged with the chickens. Owen contended, with equal fervor, that the dodo was a type of vulture. The complexity of the matter of properly classifying the dodo is illustrated in Owen's paper "Observations on the Dodo, *Didus ineptus*," first read before a public session of the British Association for the Advancement of Science in July 1846.35

New deposits of bones of a presumably extinct "new genus of gigantic wingless Birds," including "five species, one with the astonishing stature of ten feet," had recently been dug up in New Zealand, and the kiwi *Apteryx* entered the academic scene, with Owen himself as the leading expert on the new species. The study of bygone gigantic birds was one of Owen's main fascinations, triggered in part by the discovery of the colossal New Zealand moa, which he named *Dinornis* in 1843. This was the first known fossil of a bird. Then, in a 1863 publication, Owen reported on an unusual Jurassic fossil bird—the *Archaeopteryx*—which had been found in Germany. Mounting evidence that the fauna of our planet had indeed encompassed several groups of huge flightless birds, scattered in different latitudes, was giving more credence to the actual existence of the big dumb dodo.

Adding to the excitement caused by *Dinornis* and *Apteryx* was the confirmation that some sort of colossal ostrich, another extinct giant bird, had until recently roamed the plains of Madagascar. This creature, classified as *Aepyornis maximus*, stood ten feet high, weighed half a ton, and laid eggs two gallons in size; it became extinct when Europeans reached Madagascar in the sixteenth century and began hunting it and its eggs.

It is easy to imagine the effect that the mammoth shadow of *Aepyornis*'s existence and its demise could cast over nineteenth-century ornithology. Natural and man-made extinction of colossal precursors of modern species was becoming a more clear-cut and consistent phenomenon. In a paper first read at a public session in 1848 and titled *On Dinornis*, Owen cited proliferating finds of prehistoric remains of large animals in Asia, Europe, and South America, as well as in Australia and New Zealand. Especially remarkable and conclusive was

> the repeated discovery in the fluviatile deposits of New Zealand of the remains of gigantic forms of bird, allied to those small species which still exist there and there alone. This conformity of geographical localization of the extinct gigantic with the existing smaller birds of New Zealand is more striking when we remember that they constituted the highest representatives of the warm-blooded terrestrial

animals in the island, which, prior to the advent of Man, appears to have been destitute of any terrestrial mammals, and to owe its present examples of the class to the "faithful dog" which originally accompanied the Maori, and to the attendant herds and murine vermin [mice and rats] that have been recently introduced by European voyagers and colonists.[36]

The New Zealand findings in particular induced Owen, writing in the opening of his final *Memoir of the Dodo,* "to subjoin a few observations which I made during a recent visit to Oxford, on the famous head and foot of the dodo preserved in the Ashmolean Museum of that University."[37] Owen might have been fond of grandiose disputes, but, in this particular case, he did not choose to classify the dodo as a form of vulture simply for the sake of flexing his muscles in a difference with his old friend Cuvier. Rather, his measurements and comparisons led Owen to focus on the dodo's similarity with vultures. Owen's plight is a perfect illustration of the great challenge presented to his contemporary dodologists, one of which would take them to epic heights of descriptive anatomy. With the dodo vanished from earth and only a half-dozen allegorical paintings and a random bunch of naïve sketches and prints to rely on, they had to turn time and again to a skull and a foot, then beg the bones to talk.

Owen assessed the bones with eloquence and precision. In a clear proof of sincerity, he actually started with those details that seemed to set the dodo *apart* from vultures. After a long and careful examination of the various bones of the skull, Owen then turned to the foot. It was here that Owen found a number of similarities between dodos and vultures. These similarities had a double effect, not only making the dodo seem close to vultures, but also distancing it from eagles, another suggested family for the crazy bird. All things considered, Owen concluded that the bones had a plausible story to tell:

> Upon the whole, then, the Raptorial character [here referring
> to vultures exclusively] prevails most in the structure of the
> foot, as in the general form of the beak of the Dodo, and the
> present limited amount of our anatomical knowledge of the
> extinct terrestrial Bird of the Mauritius supports the
> conclusion that it is an extremely modified form of the
> Raptorial order. Devoid of the power of flight, it could have
> had small chance of obtaining food by preying upon the
> members of its own class; and if it did not exclusively subsist
> on dead and decaying organized matter, it most probably
> restricted its attacks to the class of Reptiles and to the littoral
> fishes, crustacea, etc, which its well-developed back-toe and
> claw would enable it to seize and hold with a firm grip.[38]

This said, Owen acknowledged that a full decoding of the language of the bones was far from achieved. He asserted that, if Mauritius and Rodrigues were searched as thoroughly as New Zealand had been for extinct wingless birds, then "our knowledge of the nature and affinities of the Dodo" would be significantly advanced.

But, despite his cautious approach, Owen simply had it wrong. It is true that the dodo's big hooked beak bore some resemblance to the vulture's bill, but it was otherwise unbelievable that the plump, waddling Dodo had anything to do with the lean, soaring scavenger. And, as much as some details of their feet suggested the feet of vultures, a careful survey of the books by all authors who had seen live dodos, whatever their discrepancies, shows unanimity in one important detail—dodos were vegetarian, subsisting on a diet poles apart from that of the meat-eating vultures.

―――――――――――

Ironically, the fossil record that first led Owen into his taxonomical probe of the dodo proved treacherous. Later studies carried out on his beloved *Dinornis* show Owen had analyzed the moa upside down. The authors of a study based on a 1954 reexamination of the fossil record observed that the dorsal side of the animal had been mistaken for the ventral side in earlier studies. Moreover, Owen had overlooked two key features: the breastbone, which was flat, proved that the moa could glide but not fly; and the natural cast of the brain case was like that of a reptile.

If the dodo was neither a chicken nor a vulture, what exactly was it? As the hold of Cuvier and his followers on taxonomy was loosened, other possibilities emerged. Over the years it would be related to penguins, to snipes, then to the ibis and the crane. But the best alternative possibility was advanced early on, when the American naturalist Samuel Cabot, in his 1847 paper "The Dodo (*Didus ineptus*): A Rasorial and Not Rapacious Bird," noted that "Cuvier took one side of the question and Prof. Owen takes the other," and proceeded to demonstrate that neither man was right.39

"After an examination of a head, sternum and humerus," wrote Cabot, "discovered under a bed of lava in Ile de France [Mauritius], Cuvier says that *they left no doubt in my mind, that this huge bird was one of the gallinaceous tribe.*" According to Cabot, this association was wrong. But the association with vultures made no sense, either. The bones, if looked at from a different angle, revealed a different analogy. "Mr. Owen has, I believe, examined only the head and the foot," wrote Cabot.

> He describes how the skull of the Dodo differs from that of the *Vultridae*. Now, in these very points in which Mr. Owen says this bird differs from the Order with which he connects it, it does agree with the *Columbidae*. All pigeons have the high forehead, some more than others. Then Mr. Owen omits one point, in which the Dodo differs from all rapacious birds, and indeed from all other birds, I believe, except the pigeons, and some Waders, viz., the bulging out of the lower mandible on its sides beyond the upper; we see this most strongly marked in young pigeons in the nest, at which time their general shape has a striking resemblance to that of the Dodo.40

Therefore, according to Cabot, we had to admit that both Cuvier, "the great father of science himself," and Owen, "a very great and excellent comparative anatomist," were wrong. Distracted by chickens and birds of prey, they had overlooked analogies to the pigeon. Discussing the dodo's foot, Cabot noted that:

> The articulating surface resembles that of the pigeons except in those points in which we should expect it to differ; it is more deep and strongly marked, which difference would be necessary on account of the much greater weight it has to sustain, and to the much greater importance that no dislocation should take place, the bird having no other means of locomotion. The general shape and proportions of the foot are almost the same as those of some pigeons, the toes being shorter and stouter. The claws are much like some of the ground doves, and not at all like those of any rapacious bird. The sole of the foot has none of the prominent rough callosities, which we see on the feet of all Raptores; and in this, again, resembles the pigeons. The ends of the toes have not the enlargement for the reception of the claws which we see in all rapacious birds, but are precisely like those of the pigeons.[41]

Cabot backed his opinion with the accounts of van Neck and van Warwijk of de Bry (a seventeenth-century Dutch voyager), Clusius, Sir Thomas Herbert, and Bontius—all early fellow travelers in the field of dodology. All of their descriptions contained hints that the dodo might have been a sort of pigeon. After carefully studying the data, Cabot came to several conclusions. First, the flesh of the dodo was acceptable to eat, which would not be the case if it were a vulture. Second, the fatness of the dodo deprived it of the ability to eat animal flesh. Third, dodos had gizzards, which no raptor had.

Finally, "I think it very clear that the Dodo was a gigantic pigeon, and that in its general shape, feathering, etc., resembles more strongly the young than the adult pigeon. We may perhaps be allowed to surmise that it properly belongs to an earlier epoch than the present, and has become extinct because its time was run."42

Cabot's position was reinforced in 1848, when Strickland published *The Dodo and Its Kindred*. No skeletons were available yet, but Strickland had made the most of studying Roelandt Savery's paintings. He came to firmly believe that the dodo was actually a modified giant dove. The original stock had settled on the Mascarenes a long, long time ago. In adapting to the islands' habitats, the animals had grown heavy and finally lost the ability to fly, which they didn't need since there were no predators to which they were vulnerable. They had lost all notion of fear and quietly survived while millennia passed, and the three cousins evolved into a degree of specialization that would doom them as soon as any dangerous animal set foot on shore.

Strickland's concept of the dumb, fat dodos as gentle elegant doves raised eyebrows and drew laughter. Strickland himself died in a railway accident in 1853, before he could see his theory vindicated. Soon after his death, however, explorers on the Pacific island of Samoa discovered a large, powerful bird with a thick, hooked beak. It seemed like a huge dove, but its beak looked like a dodo's. Only a study published in the March 1, 2002, issue of *Science* eventually proved that the tooth-billed pigeon of Samoa is indeed the intermediary link between the pigeons of Europe and the crazy birds of the Mascarenes.43

———————

The efforts of the Oxford zoologists to collect and analyze all outstanding data on the dodo succeeded in establishing the reality of a once-real bird. These efforts also generated interest in the once-forgotten bird on its native soil of Mauritius, now a British colony. Amateur naturalists formed the Society of Natural History and went looking for dodo bones. They were unsuccessful until one of them sailed to Rodrigues and found a good number of large, unfamiliar bones in caves. These were sent to Europe, and after further study were proclaimed to be remains of Leguat's solitaire.

This finding stimulated the Mauritian naturalists to new efforts, this time led by a schoolteacher named George Clark. They had no success until Clark had an insight. The soil of Mauritius, he observed, was not suitable for the deposit of fossil remains because it was mostly thick clay or volcanic lava, which meant that heavy tropical rains striking the hard ground would wash any bones away before they could safely be buried in the earth.

Where to look for bones, then? Under water, Clark reasoned. Three rivers of Mauritius met and ran into the sea near the town of Mahebourg, forming a muddy, marshy delta, a possible site for the elusive bones. In 1863, Clark hired laborers, who began excavating the marshes in an area known locally as the *Mare aux Songes* (Pond of Dreams). At first, Clark and his men found only a few specimens of bones; then he, as he would recount, "thought of cutting away a mass of floating herbage nearly two feet in thickness, which covered the deepest part of the marsh. In the mud under this, I was rewarded of finding the bones of many Dodos."[44]

Indeed, the bones were in such great abundance that eventually they could be assembled into complete specimens and, after first being sent to England, were shipped to museums all over the world, among them the American Museum of Natural History in New York and the Smithsonian Institution in Washington. Clark's full report on his discovery was published in 1865, which coincidentally was the year of the first printing of *Alice's Adventures in Wonderland*, from which the general public got its first introduction to the dodo.

In his account of the discovery, Clark noted: "All the specimens appear to have belonged to adult birds; and none bear any marks of having been cut or gnawed, or of the action of fire. This leads me to believe that all the Dodos of which the relics were found here were denizens either of this marsh or

its immediate neighborhood; that they all died a natural
death; and that they were very numerous in Mauritius."

The location of the bones led Clark to extrapolate more
details of the dodo's lifestyle—details that fit neatly with the
accounts of seventeenth-century sailors:

> The Mare aux Songes and the lands around it were covered
> with thick forests at the beginning of the present century:
> now not a tree remains. From its sheltered position and the
> perennial springs which flow in it, it must have afforded a
> suitable resort for birds of all kinds, and was probably a
> favorite abode of Dodos and marsh birds.[45]

This reasoning gave Clark a clear idea of just how much
the original habitat of Mauritius had been degraded, leading
to firmer conclusions about the dodo's diet. Even though no
one alive on the island had the faintest idea of the bird whose
bones were buried in the mud of the *Mare aux Songes*,

> Aged persons who have passed their lives in the woods have
> assured me that there was formerly a sufficiency of wild fruits
> to maintain any number of birds large enough to eat them,
> and that there was such a succession of them as would have
> sufficed for the whole year. I think it likely that seeds of
> several species, notwithstanding their hardness, may have
> been eaten by birds whose digestive powers we may imagine
> to have been equal to those of the Ostrich.[46]

Clark's findings filled in the missing pieces of the cen-
turies-old puzzle of the dodo. A family of related birds had
indeed existed in the Mascarenes, until they suffered a head-
on collision with humans in the seventeenth century and disappeared forever. The gray dodo of Mauritius had been the
first to succumb, followed by his white cousin on Réunion,
and eventually by the brown solitaire of Rodrigues. A careful
analysis of bones made it clear that these doomed creatures
had been every bit as bizarre as the descriptions in the old
sailors' tales.

Now it was beyond dispute that millions of years ago, a side branch of the ancestors of modern doves flew over the Indian Ocean and settled in the isolated Mascarenes. Adapting to life on the ground, they grew bigger and clumsier to the point where they could hardly get around at all. And when the birds finally had to face the intrusion of man, they were no longer able to adopt to the presence of such a tenacious predator. The dodo skeletons dug out by George Clark in the 1860s proved Strickland entirely right. The dodos were indeed a kind of dove, the most robust and awkward of them all. Slowly molded by their habitat, they became all the more vulnerable—specialized to the point of being an ideal candidate for speedy extinction.

1 Hugh Edwin Strickland, *The Dodo and Its Kindred; or, The History, Affinities, and Osteology of the Dodo, Solitaire, and Other Extinct Birds of the Islands Mauritius, Rodriguez, and Bourbon*, London: Reeve, Benham, and Reeve, 1848.
2 *Ibid.*
3 *Ibid.*
4 *Ibid.*
5 Jakob de Bontius (1658) and James Bontius (trans.), *De medicina Indorum: An Account of the Diseases, Natural History and Medicine of the East Indies*, London: T. Noteman, 1769. As quoted in Hugh Edwin Strickland, *The Dodo and Its Kindred; or, The History, Affinities, and Osteology of the Dodo, Solitaire, and Other Extinct Birds of the Islands Mauritius, Rodriguez, and Bourbon*, London: Reeve, Benham, and Reeve, 1848.
6 Walker Charleton, *Onomasticom zoikon: plerorumque animalium differentias & nomina propria pluribus linguis exponens: cui accedunt mantissa anatomica, et quaedam de variis fossilium generibus*, London: Jacobum Allestry, 1668.
7 Victor Proetz, "Diary of the Dodo," *Museum News* 42 (1964), n. 5: 25.

8 Stephen Jay Gould, "The Dodo in the Caucus Race," *Natural History* 105 (1996), n. 11: 22.

9 George-Louis Leclerc, Comte de Buffon, "Empire de l'homme sur les animaux," in: *Les animaux*, Paris: Jean Grassin, 1980.

10 George-Louis Leclerc, Comte de Buffon, *Buffon's Natural History: Containing the Theory of Earth, a General History of Man, of the Brute Creation, and of Vegetables, Minerals, etc, etc*, London: H. D. Symonds, 1797.

11 *Ibid.*

12 *Ibid.*

13 *Ibid.*

14 Stephen Jay Gould, "The Dodo in the Caucus Race," *Natural History* 105 (1996), n. 11: 22.

15 George-Louis Leclerc, Comte de Buffon, *Buffon's Natural History: Containing the Theory of Earth, a General History of Man, of the Brute Creation, and of Vegetables, Minerals, etc, etc*, London: H. D. Symonds, 1797.

16 Quentin Keynes, "Mauritius: Island of the Dodo," *National Geographic* 109 (1956), n. 190: 77.

17 William John Broderip, "Notice of an Original Painting, Including a Figure of the Dodo, in the Collection of the Duke of Northumberland, at Sion House," *Ann. & Mag. Natural History* 2 (1855), n. 15: 143.

18 Hugh Edwin Strickland, *The Dodo and Its Kindred; or, The History, Affinities, and Osteology of the Dodo, Solitaire, and Other Extinct Birds of the Islands Mauritius, Rodriguez, and Bourbon*, London: Reeve, Benham, and Reeve, 1848.

19 *Ibid.*

20 *Ibid.*

21 *Ibid.*

22 Morel (1798), "Les oiseaux monstrueux nommés Dronte, Dodo, Cygne Capuchoné, Solitaire et oieau de Nazare," in: Hugh Edwin Strickland, *The Dodo and Its Kindred; or, The History, Affinities, and Osteology of the Dodo, Solitaire, and Other Extinct Birds of the Islands Mauritius, Rodriguez, and Bourbon*, London: Reeve, Benham, and Reeve, 1848.

23 Hugh Edwin Strickland, *The Dodo and Its Kindred; or, The History, Affinities, and Osteology of the Dodo, Solitaire, and Other Extinct Birds of the Islands Mauritius, Rodriguez, and Bourbon*, London: Reeve, Benham, and Reeve, 1848.

24 C. Wesley Williams, *Dictionary of Scientific Biography*,
    New York: Scribners, 1980.

25 As quoted in: Stephen Jay Gould, "The Dodo in the
    Caucus Race," *Natural History* 105 (1996), n. 11: 22.

26 As quoted in: C. Wesley Williams, *Dictionary of
    Scientific Biography*, New York: Scribners, 1980.

27 Hugh Edwin Strickland, *The Dodo and Its Kindred; or,
    The History, Affinities, and Osteology of the Dodo, Soli-
    taire, and Other Extinct Birds of the Islands Mauritius,
    Rodriguez, and Bourbon*, London: Reeve, Benham, and
    Reeve, 1848.

28 *Ibid.*

29 Richard Owen, *Memoir of the Dodo* (Didus ineptus,
    Linn.), London: Taylor and Francis, 1866.

30 Hugh Edwin Strickland, *The Dodo and Its Kindred; or,
    The History, Affinities, and Osteology of the Dodo, Soli-
    taire, and Other Extinct Birds of the Islands Mauritius,
    Rodriguez, and Bourbon*, London: Reeve, Benham, and
    Reeve, 1848.

31 Richard Owen, *Memoir of the Dodo* (Didus ineptus,
    Linn.), London: Taylor and Francis, 1866.

32 *Ibid.*

33 Hugh Edwin Strickland, *The Dodo and Its Kindred; or,
    The History, Affinities, and Osteology of the Dodo, Soli-
    taire, and Other Extinct Birds of the Islands Mauritius,
    Rodriguez, and Bourbon*, London: Reeve, Benham, and
    Reeve, 1848.

34 Richard Owen, *Memoir of the Dodo* (Didus ineptus,
    Linn.), London: Taylor and Francis, 1866.

35 Richard Owen, "On Dinornis," *Transactions of the
    Zoological Society of London* III (1839–1848): 235, 307,
    345.

36 *Ibid.*

37 Richard Owen, *Memoir of the Dodo* (Didus ineptus,
    Linn.), London: Taylor and Francis, 1866.

38 Richard Owen, "On Dinornis," *Transactions of the
    Zoological Society of London* III (1839–1848): 235, 307,
    345.

39 Samuel Cabot, "The Dodo (*Didus ineptus*): A Rasorial
    and Not Rapacious Bird," *The Boston Journal of Natural
    History* 5 (1847): 490.

40 *Ibid.*

41 *Ibid.*

42 *Ibid.*

43 George Clark, "Account of the Late Discovery of Dodo's Remains in the Island of Mauritius," *Ibis* 2 (1865): 141–146.
44 Beth Shapiro *et al.*, "Flight of the Dodo," *Science* 295 (2002): 1683.
45 George Clark, "Account of the Late Discovery of Dodo's Remains in the Island of Mauritius," *Ibis* 2 (1865): 141–146.
46 *Ibid.*

Chapter 7

An Enduring Legacy

WHAT IS THE MOST READ, the most translated, the most reprinted, the most circulated book in the Western world?

As just about everybody knows, it is the Bible.

Now how about the second most read, most translated, most reprinted, most circulated book in the world? Maybe you would be surprised.

Robinson Crusoe, by Daniel Defoe, first published in London in 1719 (around the time the author celebrated his sixtieth birthday), was reprinted six times while the Defoe was still alive, and went ahead to become the Western best-seller of all time. By now, it has been translated into all the known languages in the world. And yes, it ranks second only to the Bible in the list of the most-read texts on the planet.[1]

Why are we so enthralled by stories consisting of lonely castaways starting civilization from scratch on some faraway island? Perhaps it's the exhilarating sense of possibility. Just imagine the tremendous feeling of excitement about living out one's dreams in Eden-like tropical surroundings that must have been in the air during the seventeenth and eighteenth centuries. Around that time, these island paradises seemed to truly exist, and reports of men stranded on them were spreading across Europe. Fact and fiction were so tightly mingled in those stories that Defoe was accused of plagiarism soon after *Robinson Crusoe* was published. When the literature of the time is surveyed, the accusation seems warranted. Seven years

earlier, a strikingly similar book had been written by Woodes
Rodgers. In it, he told the (supposedly) true story of Alexander Selkirk. Selkirk, a Scottish sailor, was abandoned on the
small island of Juan Fernandez in the Pacific Ocean off the
coast of Chile, where he survived for many years until he was
retrieved by a ship that took him back to England.

At least three similar Portuguese books were published
before *Robinson Crusoe*: *Décadas da Ásia* by João de Barros,
Lendas da Índia by Gaspar Correia, and *Descoberta e História da Conquista da Índia, na Oficina de Pedro Ferreira
para a Casa Real* by Fernão Lopes de Castanheda. They tell
all the story of Fernão Lopes, who joined forces with the
Moors and fought against Portugal during the siege of Goa in
the early sixteenth century. Imprisoned after the powerful
Portuguese viceroy Afonso de Albuquerque captured the city,
Lopes had his nose, right hand, and left thumb cut off as pun-
ishment in view of the whole population. During the event,
"Boys tore away at all the hairs in his head and beard, and covered him with mud and dung from the pigs, that was made
ready for the occasion."[2] Miraculously, Lopes survived the
extensive bleeding and was sent to Lisbon in chains on a
departing caravel. During the stopover at St. Helena, he managed to escape from the ship, and proceeded to survive by
hiding in the woods for many years, first by himself, and then
in the company of a rooster and a runaway mulatto slave.

Just like the fictional Crusoe, Lopes eventually returned to
Europe, and there fulfilled a dream. Crusoe dreamt of distribut-
ing riches to his entire family, while Lopes wanted to go to
Rome and beg forgiveness from the Pope. Next, just like Crusoe, Lopes returned to his island to accomplish one last goal.
Crusoe wanted to help the natives, who had meanwhile prolifer-
ated, while Lopes chose to spend the rest of his life in penitence.

The line between fact and fiction becomes even more
blurred when we consider that several contemporaries of
Defoe, such as Charles Gildon, were quick to classify *Robinson Crusoe* as some sort of dreamed-of autobiography. In real

life, Defoe was arrested several times, both for financial and
political reasons. During one of those arrests, both his ears
were cut off, and he had to endure the humiliation of being
exposed in this state at Cornhill, in the center of London.
Bankruptcy, shame, and pain became his constant companions. Dreaming of himself as alone and surviving on a island
thanks to his intelligence and willpower may have come as a
natural escape to Defoe.

———————

While Defoe may have only fantasized about starting anew on
an island, this dream was a reality for others. We should
remember at this point that the pious François Leguat ended
up memorialized in history as "the French Robinson Crusoe." Just seven years before *Robinson Crusoe* was published,
Leguat and his companions set sail for their impossible Eden.
Leguat and his companions were not castaways, but were the
beneficiaries of Marquis Henri du Quesnes. The French
nobleman was sympathetic to the French Protestants' plight
and actually dreamed up a scheme to send them to an Eden in
order to start society anew. The pamphlet that du Quesnes
published to recruit Protestants for the voyage included a
description of Réunion as the land of milk and honey on
earth. These romantic notions of the new worlds that could
be created clearly show us how much European thinking had
changed from the days when unknown lands were populated
with Polyhistor's terrible creatures.

The Protestant Reformation and its mixed outcomes certainly encouraged dreams of starting over as far away as possible from an Old World that seemed more divided than ever.
The beautiful (and bountiful) islands revealed by the expansion of the maritime trade routes must have appeared as natural settings for these dreams, being doubly blessed with isolation and smallness. Writing in the sixteenth century, Thomas
More installed his *Utopia* on just such an island. Like *Robin-*

son Crusoe, *Utopia* was a delicate combination of fact and fiction, in this case for the sake of satire as a means to change a rotting world. Ultimately, though, *Utopia* was an effective hoax, complete with an ingenious woodcarving representing the dreamland of harmony and a sample of its inhabitants' dialect. When it was published, many readers took the work as fact, and one missionary even planned to set sail for the mythical place in order to convert the Utopians to Christianity.

But from where did this vision of Utopia come? In 1515, More was sent by Henry VIII to an embassy in Brugge. From Brugge he moved on to Antwerp, where Giles, the city's clerk, introduced him to a bearded, weather-worn Portuguese sailor, the wanderer Rafael Hythlodaye. Hythlodaye claimed that he had sailed with Amerigo Vespucci's fleet, and had traveled around the globe six years before Magellan's voyage. During his explorations somewhere in the waters of the New World, he said he had landed on a happy island whose inhabitants had found the key to solving all the problems that plagued Europe at the time.

Hythlodaye described the secret of their happiness to More, claiming that:

> Among the Utopians . . . all things being common, every man
> hath abundance of everything. . . . I hold well with Plato . . .
> that all men should have and enjoy equal portions of health
> and commodities. . . . For where every man, under certain
> titles and pretenses, draweth and plucketh to himself as much
> as he can, so that a few divide among themselves all the whole
> riches . . . there to the residue is left lack and poverty.[3]

In the Utopia that More envisioned, each man takes his product to the common store, and receives from it according to his needs. No one asks for more than enough, for security from want forestalls greed. There is no money in Utopia, no buying cheap and selling dear: the evils of dealing, cheating, and fighting over property are unknown. Every family engages in both agriculture and industry—men and women alike. In order to ensure adequate production, six hours of work per day are required of each adult. There are laws in Utopia, but they are simple and few. If a law is violated, the citizen is expected to plead his own case, and no lawyers are allowed. While some of More's visions for Utopia may seem repressive or hopelessly naïve to twenty-first century readers, at the time of its publication, his ideas appealed to those saddened by the strongly divided, spiritually tortured, and socially corrupt England of Henry VIII.

━━━━━━━━━━━

Now, if these exotic and magical islands captured the imagination of Europe from the fifteenth to the eighteenth centuries because they really existed, the fact that most of them were inhabited and colonized by the nineteenth century didn't diminish their appeal. Flowers, fruits, whole trees, embalmed rhinos, and live birds from these far-away Edens flooded the European seaports. The abundance and variety of these specimens first led to a euphoric notion of life in which all forms of Nature's self-expression were possible. The exotic flora and fauna also triggered a big push forward in understanding and classifying the living world, and eventually helped establish and develop the modern notion of taxonomy.

Through the combination of taxonomy and geology, the notion of evolution was born. Evolution implicitly brought along the notion of extinction, and one decade before Darwin published *On the Origin of the Species*, scholars like Strickland and Owen had noticed that species extinction could be caused by man, and not just by God or Nature. For both scholars, the first clear example of this newly discovered phenomenon was the dodo.

Consider the impressive dodo lore inherited by the twentieth century. The dodo was an exotic bird of extremely unlikely appearance, living on a small cluster of tropical islands, and nowhere else in the world. First described in the travel journals that Europe craved during the sixteenth and seventeenth centuries, the dodo rose to fame through the same type of real-life voyages that would lead to the fame of a fictional Crusoe. Even long after the last dodo had vanished, it lived on as its existence was furiously debated in the academic struggles surrounding taxonomy and evolution. And, in the end, the dodo was finally identified—for the first time in the long history of human ideas—as a victim of species extinction caused by human intervention. Just look at the literature. We received from our ancestors a dodo with a monumental shadow.

First of all, since it was the first creature whose disappearance triggered human awareness, the dodo reached us as a cliché for anything vanished. We may not know the whole story, but we know what "dead as a dodo" means. Whenever we say this, we are unwittingly paying tribute to important events that brought seventeenth-century Europe to Western modernity.

That the dodo has become a self-explanatory symbol is clearly illustrated in studies such as *Retreat of the Dodo: Australian Problems and Prospects in the '80s*. It is a small treatise analyzing "the eclipse of national identity," with chapters focusing on topics such as "Australia Transformed," "Change & Resistance," and "Protection as an Article of Faith." The third chapter—which addresses "the tyranny of pragmatism and adversary games and ritual confrontation"— is called "Ways of the Dodo."[4] The epigraph is from a poem by Hilaire Belloc:

> The Dodo used to walk around
> And take the sun and air
> The sun yet warms his native ground
> The Dodo is not there[5]

The book is about change, as the author indicates in the first sentence of the introduction, but of change analyzed through the basic question: "Why do things that might and should happen fail to happen?" No one needs to be told what the dodo means; the perception of the metaphor is so obvious that the author does not feel the need for an explanation.

The other immediate association with the word "dodo" today refers to all kinds of stupidity, from the noble savage to the lame duck. During the first half of the twentieth century, this association allowed the English novelist Edward Frederic Benson—son of the archbishop of Canterbury and one time teacher in Athens and Egypt, biographer of Queen Victoria and William II of Germany, listed in the *Who's Who* as a graduate of King's College and a regular at golf, tennis, and skating facilities with a membership at the Bath club—to write a long collection of novels about a character based on the long-departed dodo. The enterprise started with *Dodo*. The book's success inspired the author to continue the line with *Dodo: A Detail of the Day*, *Dodo's Daughter*, and *Dodo Wonders*. Although Benson published over 100 titles during his life, it was the dodo books that transported him to posterity. In these books, Dodo is a woman. And we immediately understand that this woman is a dodo, even if she is one in name only. In *Dodo Wonders*, Dodo is searching for a certified pure-blooded English husband for her daughter Nadine. In spite of her efforts, she is constantly embarrassing her daughter and the rest of the English-born company with one *faux pas* after another because she is not from England. For instance, consider the opening sentence of *Dodo Wonders*:

> Dodo was so much interested in what she had herself been saying, that having just lit one cigarette, she lit another at it, and now contemplated the two with a dazed expression.[6]

Or, along the same lines, one of the first things Dodo's daughter says about her mother in *Dodo's Daughter*:

> "Heredity is such nonsense," said Nadine crisply, speaking
> with that precision which the English-born never quite attain.
> "Look at me, for instance, and how nice I am, then look at
> Mama and Daddy." Esther spilt a larger quantity of camomile
> tea than usual. "You shan't say a word against Aunt Dodo,"
> she said.[7]

So now we know that poor Dodo is not only stupid, she is
not even English-born. She's Austrian, in fact, with a give-
away dumb accent to match.

It is hard to tell for how long "dodo" has been used as an
immediate analogy for "stupid." One detail we know for sure:
soon after the nineteenth-century controversy surrounding
Darwinism (see Chapter 6), the power of such an epithet was
already evident to the sharpest wit of all wits—the young student Oscar Wilde. Before he reached the status of almighty
quote dispenser for the joy of posterity, he called one of his
professors "that illiterate dodo." Wilde seems to have acci-
dentally pioneered a now blooming Western tradition.

Today, "dodo" can even function as a two-faced pejora-
tive, applying to something dumb and about to vanish. We can
find this double whammy put to excellent use in Noel
Coward's biting description of the English upper class, *Not
Yet the Dodo*, or expressed in the most subliminal manner
through the impossible pessimism of any poem contained in
Christopher Logue's *Ode to the Dodo*. In both cases, the word
"dodo" is used in the title alone and is absent from the contents, since the contents are supposed to make clear what "dodo"
stands for. Similarly, Frank Swettenham made a great use of the
word in his caustic mockery of life in colonial Mauritius. To
convey the idea that everything and everyone in the island is both
idiotic and inviable, the author simply called the then-British
colony Dodo Island. And David Quammen plays to this binary
condition with great efficiency, titling his volume on island biodi-
versity and extinction *The Song of the Dodo*.

Bearing witness to the longevity of the "extinct" connota-
tion, we can cite a 1935 pamphlet destined to direct the atten-

of the latest copyrighted works by

BRITISH AND AMERICAN AUTHORS

Vol. 209

E. F. BENSON

Dodo
Wonders

IN ONE VOLUME

WILLIAM COLLINS SONS & CO. LTD.

PUBLISHERS
9 RUE DES HIRONDELLES   BRUXELLES
ALSO AT LONDON AND GLASGOW

Price 4.50 Prix

Dodo Wonders *is one of a series of several books on a lady from Austria trying to pass herself off as really distinguished and sophisticated among the English gentry—which, of course, makes her look dumb at every turn. Predictably enough, the series was written by a golf-loving member of the Bath Club, who had garnered several entries in the English* Who's Who. (Courtesy of the President and Fellows of Harvard College.)

tion of local authorities to the probable disappearance of the wild duck population in a certain New Hampshire pond. The suggestive title *Shall Ducks Follow the Dodo?* sets the cautionary tone, with a cartoon to reinforce the point, followed by several pages in which the dodo is never mentioned again. Everybody knows it's extinct, like the dinosaurs. And everybody knows that, unlike the dinosaurs, it was eliminated because of human activity.

A similar principle applies to Vilhjalmur Stefansson's hefty volume *Adventures in Error*, published in 1936, which starts with the incisive passage, "It is said that Bacon considered all knowledge his province. But the sciences of today are so many and complex that a single Baconian view of them is no longer possible, and perversions of thought and action result because our intellectual horizon has been narrowed to a part of the field."[8] This book is yet another attempt to disclose a new method that would standardize contemporary knowledge under a single guiding light. Even as Stefansson labors toward this noble and lofty goal, he acknowledges we will soon realize that "in these fields, among others, the accepted facts of a dozen years ago have become the error folklore of today. You standardize knowledge, and while you are at the job knowledge changes."

Stefansson's third chapter is dedicated to an interesting reincarnation of previous approaches to knowledge. He asserts that "we offer proofs that exploration can remain durable after the last island has been discovered." The argument bends and turns smartly along recitations of how explorers of other ages changed our perception of both our immediate environment and the global world we live in. The title of this chapter? You guessed it. "Are Explorers to Join the Dodo?," with no further information needed on how the metaphor works.

In 1940, the English Darwin popularizer Julian Huxley collected some of his essays in the volume *Man Stands Alone*. "I write these lines in the London Zoo's basement shelter, to

In this 1935 cartoon, the dodo fulfills one of its symbolic roles.
(Illustration appears in a pamphlet produced by the National
Association of Audubon Societies, New York, 1935.)

the sound of AA guns outside, and inside the Holst quartet playing Sibelius' beautiful Voces Intimae on the wireless," reads the first sentence.9 Here, the example is even more to the point, because Huxley starts by telling us, right in the introduction, the theme of his essays: "They were written during that strange restless indecisive period during which an age was dying but most of us were refusing to face the imminence of its dissolution." The theme of the dodo starts to emerge: "If civilization is to recreate itself after the war, it can only do so on the basis of what, for want of a better word, we must call a social outlook," since "if we win, civilization will not be safe. It will only be saved if it can transform itself so as to overcome insecurity, frustration or despair." We need to adapt and change our strategies in order to survive. Or else we will go the way of the dodo—at least in the naïve, wishful thinking allowed during the brief period when the raging war's seemingly imminent destructions were raising the all-too-real prospect of soon having open ground on which to build civilization anew.

Reading Huxley now, after the war is long over, we know we did not adapt, we did not change our tactics, we are still confronting mounting pressures of frustration and despair, and we may yet go the way of the dodo. And we can perfectly understand the metaphor as soon as we spot it.

"The Way of the Dodo" is the title of Chapter 8. Addressing the extinction of species, Huxley says, "Some species are gone forever; such are the dodo and the solitaire, the quagga, the auroch, the passenger pigeon, and the great auk. They are total losses: man can destroy a species, but he cannot restore it."

Huxley holds this conclusion for a second to consider an interesting detail:

> Perhaps one should not say he cannot in most cases restore it;
> for the Germans have in the last decade produced a
> 'synthetic' auroch, a form reconstituted by crossing the most

primitive breeds of domestic cattle and selecting those types whose conformation most nearly resembles that of the original wild species. These resuscitated aurochs are said to be almost as ferocious as their prototypes. Such re-synthesis, however, is possible only with a wild species that has left domestic descendants: it would be a bold biologist who would undertake to produce a new dodo from a pigeon or to revive the quagga from the horse and zebra stock.[10]

However, aside from these opening statements, the dodo vanishes from sight, even though Huxley's writing is diversified enough to offer us insights such as "the kite was the chief scavenger of medieval London; now there are less than a dozen specimens in Britain" and "the lion used to be found in parts of Europe and ranged all over the east; now, apart from a small area in India, it is confined to Africa." The crazy bird was called upon to perform its literary duty as a self-explanatory symbol. The dodo prepared us for the essay's main points and caught our interest with the brief glimpse of its pathetic flight through human sight. Now that it has rendered us its ever-useful services, we send it back to its trademark obscurity. Symbolism is meant to help our parables, never to steal the spotlight.

———————

Speaking of symbolism, there is one last dodo issue that must be considered before we close this case.

Have you ever heard of intellectual imperialism? The dodo is a perfect parable for what this means.

The profound effects of successive rounds of European colonization on Mauritius's original plants and animals is colorfully illustrated in a passage from Sir Frank Swettenham's *Dodo Island*, published in 1912. When this gentleman takes to the description of the sights along the roads in Mauritius, we read through a long laundry list of omnipresent humans and their pets, without a single reference to the varied ecosystem of old:

You will, at first, be in constant dread of killing some one or
something. Old women of all colours and in every state of
decrepitude; dogs that are often quite unlike dogs, so
emaciated are they and so hairless; goats, children, geese,
fowls, and chickens—the road is littered with them; the
escapes are miraculous and the miracle does not always
happen. I have never seen a place so empty of cats and so full
of idiotic dogs. But if many of the dogs are idiots, a large
proportion of the fowls are determined suicides.

I mentioned the curious absence of cats, and it is
therefore needless to say that rats abound. Of course the rats
have a prescriptive right of occupation, for their numbers
surprised the Dutch when they first visited the island quite a
long time ago. There is a prejudice against doing anything to
disturb the rat, or interfere with the quiet enjoyment of his
recognized rights. The majority of the road dogs would be no
match for a good rat. So there are plenty of rats, and also
plenty of plague cases, and every one is unhappy except the
members of the Government Medical Department.

Officially, these gentlemen are at war with the rats. Their
mercenaries are good campaigners; they are not going to take
the bread out of their mouths by organizing pitched battles,
or large sweeping movements, or even by letting out a few
cats to harass the enemy. In Mauritius the conditions of life
are so unfair that one is never given a free choice between
good and bad: it is almost always a choice of evils, and the
absence of plague would be only the beginning of genteel
poverty to a number of people.[11]

Our British snob is not omitting any mention of the trees
and birds that should have been seen in place of the chickens
and old ladies simply to sharpen his merciless mockery of
colonial life, caught, *malgré lui*, in a century-old tangled web
of causes and effects. Instead, Swettenham is bearing witness
to a complex social process of collective oblivion. Upon its
extinction, the dodo didn't only disappear from sight.
Perhaps more relevant yet, it disappeared from local speech.
In part, this radical vanishing was caused by the island's con-
voluted history of conquests and settlements by different
European countries, each one with a precise strategic reason
to seek the territory, and different plans for the island's economic exploitation. Each new wave of settlers brought their

own heritage and traditions, and the culture of the island was constantly changing. Thus, when the French novelist Bernardin de Saint-Pierre, who lived on the island for three years, drew the inspiration for his celebrated idyll *Paul et Virginie* from Mauritius, he was recapturing the spirit of a period in the island's history that seemed almost forgotten even though only 60 years had gone by.

This syncopated social fabric is also clearly expressed in the island's human facade. French colonial buildings stand next to Indian temples and mosques. The year is punctuated by diverse religious festivals, such as the Christian Corpus Christi in June, the Hindu Maha Shivaratee in January, and the Tamil Cavadee in February and March. In an incisive verbal illustration of this delicate coexistence, the official language of the island is English, the predominant native language is French, and the lingua franca is Créole patois.

Repeated changes in rule and population help account for the fact that, soon after the Dutch left and the French took over in 1715, not even the older generations could remember the amazing ugly bird. Contemporary studies show that, as early as 1750, the inhabitants of Mauritius did not even remember that there had once been such a creature. When records from a colonial public dinner in 1816 are examined, we discover that, even though several people from 70 to 90 years of age were present, no one had any knowledge of the dodo from recollection or tradition. According to renowned Mauritian historian Auguste Toussaint, in the same year an English author noted that no one remembered the dodo as they once did during the era where the classic love story *Paul et Virginie* was set.[12] The absence of a shared past may also explain why the dodo is so conspicuously absent from any kind of folk tradition in any of the cultures that define modern-day Mauritius. Looking at modern sources, we can find details we can examine the *Dictionnaire des Termes Mauritiens*, where we learn that, in Mauritian French, adolescence is called *l'âge cochon*, or the bawdy years; but there is not an oral legacy about the dodo.[13] As Swettenham puts it, "The island has known many vicissitudes: it has suffered especially from that first of all natural laws, the law of change."[14]

─────────────

This historical discussion presents us with a curious geographic distortion: the dodo as we know it can hardly be conceived as a bird of Mauritius. Back when the dodo still had no name, the island gave it shelter and its environment sealed its fate by forming its final, celebrated shape. But the dodo we know took on its life far from Mauritius and became much more than simply a biological species.

The Europeans were the dodo's first known visitors. It was christened by Europeans. It was described in words and printed in woodcuts by Europeans. It was even brought live to Europe, so that more Europeans could see it and immortalize it in paintings. It was hunted and eaten by Europeans, and it was in Europe that its swift extinction first raised eyebrows. From the nineteenth century onward, it was studied and debated with fervor and passion by successive groups of Europeans. These men disagreed widely in their views concerning evolution and speciation, but they shared in common their dedication to the new science of dodology.

It was in Europe, not in Mauritius, that the dodo first appeared in children's dreams and nursery rhymes. It was in England, not in Port Louis, that an Oxford mathematician caught word of the academic dispute concerning a strange bird that had perhaps once existed, and became so enamored with it that he turned it into his own nickname. This man considered himself bumbling and ungainly, so he strongly identified with the ill-fated, ugly creature. His name was Charles Dodgson, and the story tells us that he stuttered whenever he introduced himself, in order to become "Do-Do-Dodgson." Dodo Dodgson then went ahead to re-invent his persona as Lewis Carroll, and under this new identity he created a jolly

John Tenniel's famous drawing of Alice with the Dodo, from the first edition of Alice's Adventures in Wonderland. *Most likely, the artist used as a model the dodo painting now exhibited at the British Museum.* (Lewis Carroll and John Tenniel (illus.), *Alice's Adventures in Wonderland; and, Through the Looking-Glass and What Alice Found There*, New York: Hurst, 1903.)

company of odd characters hiding from adults' sight in a hallucinated place under the ground.

Among these creatures stood a solemn bird with a walking stick, rather dignified, perennially pensive, faintly absurd, and fond of words with a large number of syllables. A number of animals and a curious little girl had just swum to shore after crossing the immense pond created by the tears the little girl shed when she was nine feet tall. Now they badly need to get dry. To achieve this goal, the bird suggests a strange race inside a circle in the sand (although he explains that the shape doesn't really matter), with no clear departure and no defined finish line. Half an hour later, with everybody dried off and now panting and sweating, it declares the race to be over, and then announces that all have won and must have prizes. The contents of the little girl's pocket become the prizes, but in the end, there is nothing left for her. The bird asks her to check her pocket again. The girl produces a thimble. Upon the bird's instruction, she hands it the thimble. The bird gives it back to her with a speech, "We beg your acceptance of this elegant thimble."[15] There is much applause. The girl thinks the whole thing is very absurd, but they all look so grave that she doesn't dare to laugh. In John Tenniel's drawing illustrating this famous moment in English literature, the solemn bird was immortalized as a faithful copy of one of the dodo paintings produced under Rudolf's auspices, now displayed at the British Museum.

Alice's Adventures in Wonderland was not a product of exotic lands far from Europe. And the dodo that landed in Wonderland to become the referee of the Caucus Race didn't rise to its literary status through a counterpart in the real world, but rather through a ghost conjured up in a collective effort of recovered memory.

In our day, fulfilling the fate of its second life, the dodo is no longer confined to Wonderland. It is scattered all over European and American culture, as conspicuously as it is missing from any Mauritian folklore.

The dodo, you see, is literally the most amazing bird ever to have been born in Europe.

1 Robert Foulke, *The Sea Voyage Narrative*, New York: Twayne Publishers, 1997.
2 Fernão Lopes de Castanheda, *Descoberta e História da Conquista da Índia, na Oficina de Pedro Ferreira para a Casa Real*, Lisbon, Portugal: 1561.
3 Thomas More and S. J. Edward Surtz, S. J. & J. H. Hexter (eds.), *Utopia*, New Haven, Connecticut: Yale University Press, 1965.
4 G. O. Gutman, *Retreat of the Dodo: Australian Problems and Prospects in the '80s*. Canberra, Australia: Brian Clouston, 1982.
5 *Ibid.*
6 Edward Frederic Benson, *Dodo Wonders*, New York: George H. Doran, 1921.
7 Edward Frederic Benson, *Dodo's Daughter: A Sequel to Dodo*, New York: Century, 1914.
8 Vilhjalmur Stefansson, *Adventures in Error*, New York: R. M. McBride & Company, 1936.
9 Julian Huxley, *Man Stands Alone*, London: Harper Collins, 1941.
10 *Ibid.*
11 Sir Frank Swettenham, *Dodo Island*, London: Simmer & Sons, 1912.
12 Auguste Toussaint, *History of Mauritius*, London: Macmillan Education, 1977.
13 Nadia Desmarais, *Le français à l'Ile Maurice: Dictionnaire des termes mauritiens*, Port Louis, Mauritius: Imprimerie Commerciale, 1962.
14 Sir Frank Swettenham, *Dodo Island*, London: Simmer & Sons, 1912.
15 Lewis Carroll, *Alice's Adventures in Wonderland*, New York: Hurst, 1903.

Bibliography

Addison, John, and K. Hazareesingh. *A New History of Mauritius.* London: Macmillan Publishers, 1984.

Anglicus, Bartholomaeus. "De proprietatibus rerum." In: Seymour, M. C., *et al.* (eds.). *On the Properties of Things: John Trevisa's Translation of Bartholomaeus Anglicus De proprietatibus rerum: A Critical Text.* Oxford: Clarendon Press, 1975–88.

Azuaje, Ricardo. *Autobiografía de un dodo: biodiversidad, extinction y algunas insensateces.* Caracas, Venezuela: Ediciones Angria, 1995.

Barros, João de. *Terceira Década da Ásia de Ioam de Barros: dos feytos que os Portugueses fizeram no descobrimento & conquista dos mares & terras do Oriente.* Lisbon, Portugal: Ioam Barreira, 1563.

Benedict, Burton. *Mauritius, the Problems of a Plural Society.* London: Pall Mall Press, 1965.

Benson, Edward Frederic. *Dodo.* New York: Appleton, 1893.

Benson, Edward Frederic. *Dodo: A Detail of the Day.* Chicago: Donohue, Heaneberry, 1893.

Benson, Edward Frederic. *Dodo's Daughter: A Sequel to Dodo.* New York: Century, 1914.

Benson, Edward Frederic. *Dodo Wonders.* New York: George H. Doran, 1921.

Bontekoe, Willem Ysbrandsz, and C. B. Bodde-Hodgkinson & Pieter Geyl (trans.). *Memorable Description of the East Indian Voyage, 1618–25.* London: G. Routledge & Sons, 1929.

Bowman, Larry W. *Mauritius: Democracy and Development in the Indian Ocean*. Dartmouth, New Hampshire: Westview Press, 1991.

Brito, Bernardo Gomes de. *História Trágico-Marítima*. Mem Martins, Portugal: Publicações Europa-América, 1981.

Brito, Bernardo Gomes de. *The Tragic History of the Sea*. Cambridge, U.K.: The Hakluyt Society, 1959.

Broderip, William John. "Notice of an Original Painting, Including a Figure of the Dodo, in the Collection of the Duke of Northumberland, at Sion House." *Ann. & Mag. Natural History* 2 (1855), n. 15: 143.

Cabot, Samuel. "The Dodo (*Didus ineptus*): A Rasorial and Not Rapacious Bird." *The Boston Journal of Natural History* 5 (1847): 490.

Carroll, Lewis. *Alice's Adventures in Wonderland*. New York: Hurst, 1903.

Carroll, Lewis, and John Tenniel (illus.). *Alice's Adventures in Wonderland; and, Through the Looking-Glass and What Alice Found There*. New York: Hurst, 1903.

Carvalho, Eduardo Luna de. "A anonima descoberta dos Doudos do arquipelago das Mascarenhas por navegadores portugueses (*Avis Columbiformes Raphidae*)." *Coleccão "Natura," Nova Serie*: 13, Lisbon, Portugal: Sociedade Portuguesa de Ciencias Naturais, 1989.

Castanheda, Fernão Lopes de. *Descoberta e História da Conquista da Índia, na Oficina de Pedro Ferreira para a Casa Real*. Lisbon, Portugal: 1561.

Cauche, François. *Relation du voyage que François Cauche a fait à Madagascar, isles adjacentes & coste d'Afrique, recueilly par le Sieur Morisot, avec das notes en marge*. Paris: Roche Beullet, undated.

Charleton, Walker. *Onomasticom zoikon: plerorumque animalium differentias & nomina propria pluribus linguis exponens: cui accedunt mantissa anatomica, et quaedam de variis fossilium generibus*. London: Jacobum Allestry, 1668.

Charman, Andy. *Madua ve-lamah hushmedah tsipor ha-dudo? U-sheelot aherot al baale (I Wonder the Dodo Is Dead, and Other Questions About Extinct and Endangered Animals)*. Tel Aviv, Israel: Yehoshua Orenshtain, 1997.

Clark, George. "Account of the Late Discovery of Dodo's Remains in the Island of Mauritius." *Ibis* 2 (1865): 141–146.

Coward, Noel. *Not Yet the Dodo, and Other Verses*. London: Heinemann, 1967.

Daston, Lorraine, and Katharine Park. *Wonders and the Order of Nature, 1150–1750*. New York: Zone Books, 1998.

Dauxois, Jacqueline. *L'empereur des alchimistes: Rudolphe II de Habsbourg*. Paris: J.-C. Lattes, 1996.

Desmarais, Nadia. *Le français à l'Ile Maurice: Dictionnaire des termes mauritiens*. Port Louis, Mauritius: Imprimerie Commerciale, 1962.

Erasmus, Kurt. *Roelandt Savery, sein Leben und seine Werke*. Doctoral thesis. Halle-Wittenberg, Germany: Friedrichs-Universität, 1907.

Evans, Robert, and John Weston. *Rudolf II and His World: A Study of Intellectual History 1576–1612*. London: Thames and Hudson, 1997.

Ferreira, Fernanda Durao. *As fontes portuguesas de Robinson Crusoe*. Lisbon, Portugal: Cadernos Minimal, 1996.

Foulke, Robert. *The Sea Voyage Narrative*. New York: Twayne Publishers, 1997.

Fučíková, Eliška, *et al.* (eds.). *Rudolf II and Prague: The Court and the City*. London: Thames and Hudson, 1997.

Funerary Equipment of Rudolf, King of Bohemia: The Earliest Hapsburgs at Prague Castle. Prague: Hradčany Castle Management, 1995.

Gordon, Jan B. (ed.). *Soaring with the Dodo: Essays on Lewis Carroll's Life and Art*. Charlottesville, Virginia: The University Press of Virginia, 1982.

Gould, Stephen Jay. "The Dodo in the Caucus Race." *Natural History* 105 (1996), n. 11: 22.

Gutman, G. O. *Retreat of the Dodo: Australian Problems and Prospects in the '80s*. Canberra, Australia: Brian Clouston, 1982.

Hachisuka, Masauji. *The Dodo of Mauritius*. Tokyo: 1939.

Hachisuka, Masauji. *The Dodo and Kindred Birds; or, The Extinct Birds of the Mascarene Islands.* London: H. F. & G. Witherby, 1953.

Halley, Edmond. *Miscellania curiosa.* London: 1726.

Harris, John. *Navigatorum atque Itinerarium Biboliotheca: or, A Complete Collection of Voyages and Travels: Consisting of above Four Hundred of the Most authentic Writers; Beginning with Hackluyt, Purchals, etc, in English; Ramusio in Italian; Thevenot, etc, in French; de Bry, and Grynaei Novus Orbi in Latin; the Dutch East-India Company in Dutch; and Continued with Others of Note, that Have Published Histories, Voyages, Travels or Discoveries, in the English, Latin, French, Italian, Spanish, Portuguese, German, or Dutch Tongues; Relation to Any Part of Asia, Africa, America, Europe, or the Islands Thereof, to the Present Time; with the Heads of our Most Considerable Sea-Commanders; and a Great Number of Excellent Maps of All Parts of the World, and Cuts of Most Curious Things in all Voyages; also, an Appendix, of the Remarkable Accidents at Sea; and Several of our Considerable Engagements: the Charters, Acts of Parliament, etc, About the East-India Trade, and Papers Relating to the Union of the Two Companies; throughout the Whole, All Original Papers Are Printed at Large: as the Pope's Bull, to Dispose of the West Indies to the King of Spain; Letters and Patents for Establishing Companies of Merchants; as the Ruff, East-India Companies, etc. Letters of One Great Prince or State to Another, Showing their Titles, Style, etc.; to which is Prefixed, a History of the Peopling of the Several Parts of the World, and Particularly of America, an Account of Ancient Shipping, and Its Successive Improvements; Together with the Invention and Use of the Magnet, and Its Variation, etc.* London: 1705.

Hazareesingh, K. *History of Indians in Mauritius.* London: Macmillan Education, 1976.

Hébert, François. *Le dernier chant de l'avant-dernier dodo.* Paris: Garamond, 1986.

Hendrix, Lee, *et al.* (eds.). *Nature Illuminated: Flora and Fauna from the Court of the Emperor Rudolf II.* Los Angeles, California: The J. Paul Getty Museum, 1997.

Histoire de la navigation de Jean Hugues de Linschot, hollandois, aux Indes Orientales: Contenant diverses déscriptions des lieux jusques à présent découverts par les portugais: Observations des coutumes & singularités de dela & d'autres déclarations. Amsterdam: 1619.

Holzer, Hans. *The Alchemist: The Secret Magical Life of Rudolf von Habsburg*. New York: Stein and Day, 1974.

Huxley, Julian. *Man Stands Alone*. London: Harper Collins, 1941.

Kaufmann, Thomas DaCosta. *Variations on the Imperial Theme in the Age of Maximilian II and Rudolf II*. New York: Garland Publishers, 1978.

Kaufmann, Thomas DaCosta. *The School of Prague: Painting at the Court of Rudolf II*. Chicago: The University of Chicago Press, 1988.

Kaufmann, Thomas DaCosta. *The Mastery of Nature: Aspects of Art, Science, and Humanism in the Renaissance*. Princeton, New Jersey: Princeton University Press, 1993.

Keynes, Quentin. "Mauritius: Island of the Dodo." *National Geographic* 109 (1956), n. 190: 77.

Lactantius and Michel Perrin (ed., trans.). *L'ouvrage du Dieu Createur*. Paris: Editions du Cerf, 1974.

Langton, Jane. *Dead as a Dodo: A Homer Kelly Mystery*. New York: Viking, 1996.

Leclerc , George-Louis, Comte de Buffon. *Buffon's Natural History: Containing the Theory of Earth, a General History of Man, of the Brute Creation, and of Vegetables, Minerals, etc, etc*. London: H. D. Symonds, 1797.

Leclerc , George-Louis, Comte de Buffon. "Empire de l' homme sur les animaux." In: *Les animaux*. Paris: Jean Grassin, 1980.

Leclerc , George-Louis, Comte de Buffon. *Histoire naturelle des oiseaux*. Paris: De L'Imprimerie Royale, 1770–1785.

Leguat, François, and Oliver Patsfield (ed.). *The Voyage of François Leguat*. London: The Hakluyt Society, 1891.

Linschoten, Jan Huygen van, and Arthur Coke Burnell & P. A. Tiele (eds.). *The Voyage of John Huygen van Linschoten to the East Indies: From the Old English Translation of 1578: The First Book, Containing His Description of the East*. London: The Hakluyt Society, 1885.

Logue, Christopher. *Ode to the Dodo: Poems from 1953 to 1978*. London: Cape, 1981.

Malone, Kemp. *The Dodo and the Camel: A Fable for Children Freely Told in English by Kemp Malone, after the Danish Version of Gudmund Schütte*. Baltimore, Maryland: J. H. Furst Company, 1938.

Markham, Clements Robert. *The Voyages of Sir James Lancaster, Kt., to the East Indies, with Abstracts of Journals of Voyages in the East Indies During the Seventeenth Century, Preserved in the India Office; and the Voyage of Captain John Knight (1606) to Seek the North-West Passage.* London: The Hakluyt Society, 1877.

Milet-Mureau, M. L. A. *Voyage de La Pérouse autour du monde, publié conformément au decret du 22 Avril 1791*. Paris: Imprimerie de la République, 1797.

More, Thomas, and S. J. Edward Surtz, S. J. & J. H. Hexter (eds.). *Utopia*. New Haven, Connecticut: Yale University Press, 1965.

Moree, P. J. *A Concise History of Dutch Mauritius, 1598–1711*. London: Kegan Paul International, 1998.

Müllenmeister, Kurt J. *Roelant Savery, Korterikj 1576–1639 Utrecht, Hofmaler Kaiser Rudolf II. in Prag: Die Gemälde mit kritischem Œuvrekatalog*. Freren, Germany: Luca Verlag, 1988.

Mundy, Peter. *The Travels of Peter Mundy*. London: The Hakluyt Society, 1919.

Mungur, Bhurdwaz. *An Invitation to the Charms of Mauritian Localities: A Survey of Names and Attractions of Places in Mauritius*. Vacoas, Mauritius: ELP Ltee, 1993.

Arthur Percival Newton (ed.). *Travel and Travellers of the Middle Ages*. London: K. Paul, Trench, Trubner & Co., Ltd.; New York: A. A. Knopf, 1926.

Oudemans, Anthonie Cornelis. "On the Dodo." *Ibis* 10 (1918), n. 6: 316.

Owen, Richard. "On Dinornis." *Transactions of the Zoological Society of London* III (1839–1848): 235, 307, 345.

Owen, Richard. *Memoir of the Dodo* (Didus ineptus, Linn.). London: Taylor and Francis, 1866.

Palmitessa, James R. "Material Culture & Daily Life in the New City of Prague in the Age of Rudolf II." *Medium Aevum Quotidianum*, 1997.

Pingré, Alexandre Guy, and J. Alby & M. Serviable (ed.). *Courser Vénus: Voyage scientifique à l'île Rodrigues 1761, fragments du journal de voyage de l'abbé Pingré.* Saint-Denis de La Réunion: ARS Terres Créoles, 1993.

Pitot, Albert. *T'Eylandt Mauritius, esquisses historiques (1598–1710): Procedees d'une notice sur la decouverte des Mascareignes et suivies d'une monographie du dodo, des solitaires de Rodrigue et de Bourbon et de l'oiseau bleu.* Port Louis, Mauritius: Coignet Freres, 1905.

Prévost, Abbé. *Histoire générale des voyages; ou Nouvelle collection de toutes relations de voyages par terre, qui ont été publiées jusqu'à présent dans les différentes langues de toutes les nations connues: Contenant ce qu'il y ba de plus remarquable, de plus utile, et de mieux avéré dans les pays où les voyageurs ont pénétré, touchant leur situation, leur étendue, leurs limites, leurs divisions, leur climat, leur térritoir, leur productions, leurs lacs, leurs rivières, leurs montagnes, leurs mines, leurs cités & leurs principales villes, leurs ports, leurs rades, leurs édifices, etc. Avec les moeurs et les usages des habitants, leur réligion, leur governement, leurs arts et leurs sciences, leur commerce et leurs manufactures; pour pormer un système complet d'histoire et de géographie moderne, qui répresentera l'état actuel de toutes les nations: Enrichi de cartes géographiques nouvellement composées pour les observations les plus authentiques, et des plans et des perspectives; de figures d'animaux, de végétaux, habitats, antiquités, etc.* Paris: 1750.

Proetz, Victor. "Diary of the Dodo." *Museum News* 42 (1964), n. 5: 25.

Quammen, David. *The Song of the Dodo.* New York: Simon & Schuster, 1996.

Quesne, Henri du. *Recueil de quelques memoires servans d'instruction pour l'etablissement de l'isle d'Eden.* Amsterdam: H. Desbordes, 1689.

Riviere, Lindsay. *Historical Dictionary of Mauritius (African Historical Dictionaries* 34). Metuchen, New Jersey: Scarecrow Press, 1982.

Roberts, Katerina. *Visitors' Guide: Mauritius, Rodrigues & Réunion.* Ashbourne, U.K.: Hunter, 1992.

Roque, Jean de la. *Voyage de l'Arabir heureuse: Par l'Ocean oriental, & le detroit de la mer Rouge, fait par les françois pour la première fois, dans les années 1708, 1709 & 1709; avec la relation particulière d'un voyage fait du port de Moka à la cour du roi d'Yemen, dans la seconde expédition des années 1711, 1712 & 1713; un mémoire concernant l'arbre & le fruit du café, dressé sur les observations de ceux qui ont fait ce dernier voyage; et un traité historique de l'origine & du progrés du café, tant dans l'Asie que dans l'Europe; de son introduction en France, & de l'établissement de son usage à Paris*. Paris: A. Cailleau, 1716.

Salvadori, B., Florio, L., and Cozzaglio, P. *Les animaux que disparaissent*. Paris: Bordas, 1975.

Savery, Alfred Williams. *A Genealogical and Biographical Record of the Savery Families*. Boston, Massachusetts: The Collins Press, 1893.

Selvon, Sydney. *Historical Dictionary of Mauritius (African Historical Dictionaries* 49). Metuchen, New Jersey: Scarecrow Press, 1991.

Selys-Longchamps, Michel-Edmond, baron de. "Résumé concernant la classification des oiseaus brévipennes mentionnés dans l'ouvrage de M. Strickland sur le Dodo." *Review of Zoology* 11 (1848): 292.

Selys-Longchamps, Michel-Edmond, baron de. *Sur la classification des oiseaux depuis Linné*. Brussels: F. Haeyez, 1879.

Shall Ducks Follow the Dodo? New York: The National Audubon Society, 1935.

Shapiro, Beth, *et al.* "Flight of the Dodo." *Science* 295 (2002): 1683.

Silverberg, Robert. *The Auk, the Dodo, and the Oryx: Vanished and Vanishing Creatures*. New York: Thomas Y. Crowell Company, 1967.

Solinus, C. Julius, and Arthur Golding (trans.). *The Worthie Work of Iulius Solinus Polyhistor: Contayning Many Noble Actions of Humaine Creatures, with the Secretes of Nature in Beastes, Fyshes, Foules, and Serpents: Trees, Plants, and the Vertue of Precious Stones: With Diuers Countryes, Citties and People: Verie Pleasant and Full of Recreation for All Sorts of People*. London: Printed at I. Charlewoode for Thomas Hacket, 1587.

Spicer-Durham, Joaneath Ann. *The Drawings of Roelandt Savery*. Doctoral thesis. New Haven, Connecticut: Yale University, 1986.

Stefansson, Vilhjalmur. *Adventures in Error*. New York: R. M. McBride & Company, 1936.

Storey, William K. *Science and Power in Colonial Mauritius*. Rochester, New York: The University of Rochester Press, 1997.

Strickland, Hugh Edwin. *The Dodo and Its Kindred; or, The History, Affinities, and Osteology of the Dodo, Solitaire, and Other Extinct Birds of the Islands Mauritius, Rodriguez, and Bourbon*. London: Reeve, Benham, and Reeve, 1848.

Swettenham, Sir Frank. *Dodo Island*. London: Simmer & Sons, 1912.

Toussaint, Auguste. *Une cité tropicale, Port-Louis de l'Île Maurice*. Paris: Presses Universitaires de France, 1966.

Toussaint, Auguste. *History of Mauritius*. London: Macmillan Education, 1977.

Vurm, Robert B., and Helena Baker (trans.). *Rudolf II and His Prague: Mysteries and Curiosities of Rudolfine Prague: Prague Between the Period 1550–1650*. Prague: Robert B. Vurm, 1997.

Williams, C. Wesley. *Dictionary of Scientific Biography*. New York: Scribners, 1980.

Index

Page references in *italics* refer to illustrations and to information in captions

209